I0010542

Microservices with Go

Building scalable and reliable microservices with Go

Alexander Shuiskov

BIRMINGHAM—MUMBAI

Microservices with Go

Copyright © 2022 Packt Publishing

All rights reserved. No part of this book may be reproduced, stored in a retrieval system, or transmitted in any form or by any means, without the prior written permission of the publisher, except in the case of brief quotations embedded in critical articles or reviews.

Every effort has been made in the preparation of this book to ensure the accuracy of the information presented. However, the information contained in this book is sold without warranty, either express or implied. Neither the author, nor Packt Publishing or its dealers and distributors, will be held liable for any damages caused or alleged to have been caused directly or indirectly by this book.

Packt Publishing has endeavored to provide trademark information about all of the companies and products mentioned in this book by the appropriate use of capitals. However, Packt Publishing cannot guarantee the accuracy of this information.

Group Product Manager: Gebin George
Publishing Product Manager: Kunal Sawant
Senior Editor: Nisha Cleetus
Content Development Editor: Yashi Gupta
Technical Editor: Pradeep Sahu
Copy Editor: Safis Editing
Project Coordinator: Deeksha Thakkar
Proofreader: Safis Editing
Indexer: Hemangini Bari
Production Designer: Roshan Kawale
Business Development Executive: Debadrita Chatterjee
Marketing Coordinators: Rayyan Khan and Deepak Kumar

First published: October 2022
Production reference: 2081122

Published by Packt Publishing Ltd.
Livery Place
35 Livery Street
Birmingham
B3 2PB, UK.

ISBN 978-1-80461-700-7
www.packt.com

I dedicate this book to my family and to my source of passion, inspiration, and love, Vera

– Alexander Shuiskov

Contributors

About the author

Alexander Shuiskov is a staff software engineer at Uber working on the observability and reliability of thousands of Uber microservices. He is an expert in alerting and monitoring solutions. Before Uber, Alexander worked on microservice development at multiple tech companies, including `Booking.com` and eBay.

Alexander holds an MSc in computer science from the Ural State University, as well as a degree in economics and management.

Outside of work, Alexander loves street photography, abstract art, sound production, and many sports, including cycling and running.

I want to thank all the people who supported me throughout this journey and provided their valuable feedback: my friends, family, colleagues, and my loving partner, Vera. I'm also grateful for the chance to work on this book: at first, I thought I would never have time for it, but curiosity and passion took precedence over all uncertainty. Never say never!

About the reviewer

Sadman Khan is currently a senior engineer at Uber in the observability department. He co-authored the alerting engine that has been critical to keeping Uber reliable. He graduated from the University of Waterloo, Canada, with a bachelor's degree in computer engineering. He has previously worked at Coinbase and LinkedIn.

Table of Contents

6

Asynchronous Communication   115

7

Storing Service Data   133

8

Deployment with Kubernetes   147

Part 3: Maintenance

Preface

Since its first release, the Go programming language has gained popularity among all types of software developers. Simple language syntax, ease of use, and a rich set of libraries made Go one of the primary languages for writing different kinds of software, from small tools to large-scale systems consisting of hundreds of components.

Among the primary Go use cases is microservice development – the development of individual applications, called microservices, that can play various roles, from processing payments to storing user data. Organizing a large system as a set of microservices often brings multiple advantages, such as increasing the development and deployment speed, but also brings multiple types of challenges. Among such challenges are service discovery and communication, integration testing, and service monitoring.

In this book, we will illustrate how to implement Go microservices and establish communication between them, how to enable the deployment of individual microservices and secure their interactions, and how to store and retrieve service data and provide service APIs, enabling other applications to use our microservices. You will learn about some of the industry's best practices related to all of these topics and get a detailed overview of the possible challenges along the way, as well as the possible benefits. The knowledge that you will gain by reading this book will help you to both create new microservices and efficiently maintain the existing ones. I hope this journey will be exciting for you!

Who this book is for

This book is for all types of developers: from people interested in learning how to write microservices in Go to seasoned professionals who want to take the next step in mastering the art of writing scalable and reliable microservice-based systems. The first two parts of the book, covering the development of microservices, would be useful to developers who are just starting their experience with Go or those who are interested in the best practices of organizing the Go application code base according to the industry's best standards; the final part would be useful for all developers, even the most experienced ones, as it provides great insights on maintaining and operating microservices at scale.

What this book covers

Chapter 1, *Introduction to Microservices*, will cover the key benefits of and common issues with a microservice architecture, helping you to understand which problems microservices solve and which challenges they usually introduce. The chapter emphasizes the role of the Go programming language in microservice development and lays down the foundation for the rest of the book.

Chapter 2, Scaffolding a Go Microservice, will introduce the you to the main principles of the Go programming language and provide the most important recommendations for writing Go code. It will cover the process of setting up the right structure to organize the microservice code in Go and introduce the you to an example application consisting of three microservices. Finally, the chapter will illustrate how to scaffold the code for each of the example microservices. The example microservices implemented in this chapter are going to be used throughout the book, with each chapter adding new features to them.

Chapter 3, Service Discovery, will talk about the problem of service discovery and illustrates how different services can find each other in a microservice environment. It will cover the most popular service discovery tools and walk the you through the steps of adding service discovery logic to the example microservices from the previous chapter.

Chapter 4, Serialization, will bring us to the concept of data serialization, which is required for understanding upcoming chapters covering microservice communication. The you will be introduced to the Protocol Buffers data format, which is going to be used for encoding and decoding the data transferred between our example microservices. The chapter will provide examples of how to define serializable data types and generate code for them, and how to use the generated code in Go microservices.

Chapter 5, Synchronous Communication, will cover the topic of synchronous communication between microservices. It will illustrate how to define service APIs using the Protocol Buffers format and introduce the you to gRPC, a service communication framework. The chapter will wrap up with examples of how to implement microservice gateways and clients and perform remote calls between our microservices.

Chapter 6, Asynchronous Communication, will talk about asynchronous communication between microservices. It will introduce the you to a popular asynchronous communication tool, Apache Kafka, and provide examples of sending and receiving messages, using it for our example microservices. The chapter will wrap up with an overview of the best practices for using asynchronous communication in microservice environments.

Chapter 7, Storing Service Data, will cover the topic of persisting service data in databases. The you will learn about the common types of databases and the benefits they bring to software developers. The chapter will walk the you through the process of implementing the logic for storing service data in a MySQL database.

Chapter 8, Deployment with Kubernetes, will talk about service deployment and provide an overview of a popular deployment and orchestration platform, Kubernetes. The chapter will illustrate how to prepare service code for deployment and how to deploy it using Kubernetes. The chapter will include the best practices for deploying microservice applications.

Chapter 9, Unit and Integration Testing, will describe the common techniques of testing Go microservice code. It will cover the basics of Go unit and integration testing and demonstrate how to test the microservice code from the previous chapters. The chapter will wrap up with the industry's best practices for writing and organizing tests.

Chapter 10, Reliability Overview, will introduce the you to the topic of system reliability and describe the core principles, instruments, and industry best practices for building reliable and highly available microservices. It will illustrate how to automate service responses to various types of failures, as well as how to establish the processes for keeping service reliability under control.

Chapter 11, Collecting Service Telemetry Data, will provide a detailed overview of modern instruments and solutions for collecting service telemetry data, such as logs, metrics, and traces. The chapter will provide lots of detailed examples of collecting all different types of telemetry data and list some of the best practices for working with them.

Chapter 12, Setting up Service Alerting, will illustrate how to set up automated incident detection and notification for microservices, using the telemetry data collected in the previous chapter. It will introduce the you to a popular alerting and monitoring tool, Prometheus, and show how to set up Prometheus alerts for our example microservices.

Chapter 13, Advanced Topics, will wrap up the last part of the book and cover some of the advanced topics in microservice development, such as profiling, dashboarding, frameworks, service ownership, and security. The chapter will include some examples of setting up secure communication between Go microservices using the JWT protocol.

To get the most out of this book

I suggest you get some familiarity with Go by implementing a few applications, such as simple web services. Familiarity with a Docker tool would be a plus because we will be using it for running some of the tools that our microservices will be using. Finally, I strongly suggest implementing, running, and playing with the example microservices that we will be implementing so that all your knowledge will be cemented by practice.

Software/hardware covered in the book	Operating system requirements
Go 1.11 or above	Windows, macOS, or Linux
Docker	Windows, macOS, or Linux
grpcurl	Windows, macOS, or Linux
Kubernetes	Windows, macOS, or Linux
Prometheus	Windows, macOS, or Linux
Jaeger	Windows, macOS, or Linux
Graphviz	Windows, macOS, or Linux

If you are using the digital version of this book, we advise you to type the code yourself or access the code from the book's GitHub repository (a link is available in the next section). Doing so will help you avoid any potential errors related to the copying and pasting of code.

Download the example code files

You can download the example code files for this book from GitHub at `https://github.com/PacktPublishing/microservices-with-go`. If there's an update to the code, it will be updated in the GitHub repository.

We also have other code bundles from our rich catalog of books and videos available at `https://github.com/PacktPublishing/`. Check them out!

Download the color images

We also provide a PDF file that has color images of the screenshots and diagrams used in this book.

You can download it here: `https://packt.link/1fb2C`.

Conventions used

There are a number of text conventions used throughout this book.

`Code in text`: Indicates code words in text, database table names, folder names, filenames, file extensions, pathnames, dummy URLs, user input, and Twitter handles. Here is an example: "Mount the downloaded `WebStorm-10*.dmg` disk image file as another disk in your system."

A block of code is set as follows:

```
package main

import (
    "encoding/json"
    "fmt"
    "os"
    "time"

    "github.com/confluentinc/confluent-kafka-go/kafka"
    "movieexample.com/rating/pkg/model"
)
```

When we wish to draw your attention to a particular part of a code block, the relevant lines or items are set in bold:

```
    if err := p.Produce(&kafka.Message{
  TopicPartition: kafka.TopicPartition{Topic: &topic, Partition:
  kafka.PartitionAny},
  Value:              []byte(encodedEvent),
  }, nil); err != nil {
        return err
    }
    return nil
```

Any command-line input or output is written as follows:

```
  mysql movieexample -h localhost -P 3306 --protocol=tcp -u root
  -p
```

Bold: Indicates a new term, an important word, or words that you see on screen. For instance, words in menus or dialog boxes appear in **bold**. Here is an example: "Select **System info** from the **Administration** panel."

> **Tips or important notes**
> Appear like this.

Get in touch

Feedback from our readers is always welcome.

General feedback: If you have questions about any aspect of this book, email us at customercare@packtpub.com and mention the book title in the subject of your message.

Errata: Although we have taken every care to ensure the accuracy of our content, mistakes do happen. If you have found a mistake in this book, we would be grateful if you would report this to us. Please visit www.packtpub.com/support/errata and fill in the form.

Piracy: If you come across any illegal copies of our works in any form on the internet, we would be grateful if you would provide us with the location address or website name. Please contact us at copyright@packt.com with a link to the material.

If you are interested in becoming an author: If there is a topic that you have expertise in and you are interested in either writing or contributing to a book, please visit authors.packtpub.com.

Share Your Thoughts

Once you've read *Microservices with Go*, we'd love to hear your thoughts! Scan the QR code below to go straight to the Amazon review page for this book and share your feedback.

https://packt.link/r/1804617008

Your review is important to us and the tech community and will help us make sure we're delivering excellent quality content.

Download a free PDF copy of this book

Thanks for purchasing this book!

Do you like to read on the go but are unable to carry your print books everywhere?

Is your eBook purchase not compatible with the device of your choice?

Don't worry, now with every Packt book you get a DRM-free PDF version of that book at no cost.

Read anywhere, any place, on any device. Search, copy, and paste code from your favorite technical books directly into your application.

The perks don't stop there, you can get exclusive access to discounts, newsletters, and great free content in your inbox daily

Follow these simple steps to get the benefits:

1.   Scan the QR code or visit the link below

https://packt.link/free-ebook/978-1-80461-700-7

2.   Submit your proof of purchase

3.   That's it! We'll send your free PDF and other benefits to your email directly

Part 1: Introduction

This part provides an overview of the microservice architecture model. It covers the key benefits and common issues of microservices, helping the readers to understand which problems the microservice architecture helps to solve and which issues it usually introduces. The single chapter in this part focuses on the role of the Go programming language in microservice development and lays down the foundation for the rest of the book.

This part contains the following chapter:

- *Chapter 1, Introduction to Microservices*

1

Introduction to Microservices

In this chapter, you will be introduced to **microservices** and the motivation behind them. You will understand the key benefits and common issues of the microservice architecture model and learn when to use it, as well as getting some microservice development best practices. This knowledge will help you establish a solid foundation for reading the next chapters and give you some ideas on what challenges you may face with microservices in the future.

In this chapter, we will cover the following topics:

- What is a microservice?
- Motivation to use microservices
- Pros and cons of microservices
- When to use microservice architecture
- Role of Go in microservice development

What is a microservice?

Companies worldwide have used the **microservice architecture model** so widely that it has almost become a default way of software development. Those companies have tens, hundreds, and even thousands of microservices at their disposal.

So, what exactly is the microservice model?

The microservice architecture model is organizing an application as a collection of services, called microservices, each of which is further responsible for a certain part of application logic, usually defined by a particular business capability.

As an example, consider an online marketplace application. The application may have multiple features, including search, shopping cart, payments, order history, and many more. Each feature can be so different that the code may (and, in certain cases, should) be completely independent of the rest of the application. In this example, search and payments technically have nothing in common.

In the microservice architecture model, each component would be an independent service playing its own role in the system.

Organizing each part of the application as a separate service is not necessarily a requirement. As with any architecture model or any aspect of software development, engineers need to be careful with choosing a particular approach or solution – doing an initial analysis and understanding the solution under the given conditions.

Before we proceed to the key benefits and downsides of microservices, let's see what challenges you could face when the application is not separated into multiple services.

Motivation to use microservices

In order to understand the motivation behind using the microservice architecture, it is very important to see the opposite approach – when the application is built and executed as a single program. Such applications are called **monolithic applications** or **monoliths**.

Monolithic architecture is, in most ways, the simplest model to implement since it does not involve splitting the application into multiple parts that need to coordinate with each other. This can provide you with major advantages in many cases, such as the following:

- **Small code base**: Splitting an application into multiple independent parts may significantly increase the size of the code base by introducing extra logic required for communication between the components.

- **Application logic is still loosely defined**: It is very common that parts of the application or the entire system go through major structural or logical changes, especially at the very early stages of development. This might be caused by a sudden change of requirements, priorities, changes in the business model, or a different approach to development. During the early stages of development, iterating fast can be critical not only to the development process, but also to the entire company.

- **Narrow scope of the application**: Not every service requires a decomposition and division into separate parts. Consider a service for generating random passwords – it has a single logical feature and, in most cases, it would be unnecessary to split it into multiple parts.

In all of the preceding cases, monolithic architecture would be a better fit for the application. However, at some point, services get too big to remain monolithic. Developers start experiencing the following issues:

- **Large application size and slow deployments**: At a certain point, an application can become so big that it can take minutes or even hours to build, start, or deploy.

- **Inability to deploy a particular part of the application independently**: Not being able to replace a part of a large application can easily become a bottleneck, slowing down the development and release process.

- **Higher blast radius**: If there is a bug in a certain function or library widely used across the application code, it is going to affect all parts of the system at once, potentially causing major issues.

- **Vertical scalability bottleneck**: The more logic the application has, the more resources it needs in order to run. At a certain point, it can get hard or impossible to scale the application up even further, given the possible limits on CPU and RAM.

- **Interference**: Certain parts of the application can heavily load CPU, I/O, or RAM, causing delays for the rest of the system.

- **Unwanted dependencies between components**: Having the entire application represented as a single executable leaves room for unnecessary dependencies between the components. Imagine a developer refactoring a code base, and making a change suddenly affects some important parts of the system, such as payments. Having more isolation between the components gives more protection against such issues.

- **Security**: A possible security issue in the application may result in unauthorized access to all components at once.

In addition to the possible issues we just described, different components may have different requirements, such as the following:

- **Resources and hardware requirements**: Certain components are more CPU-intensive or memory-intensive and may perform I/O operations at a higher rate. Separating such components may reduce the load on the entire system, increasing system availability and reducing latency.

- **Deployment cadence**: Some parts of the system mostly remain unchanged while others require multiple deployments per day.

- **Deployment monitoring and automated testing**: Certain components may require stricter checks and monitoring and can be subject to slower deployments due to multi-step rollouts.

- **Technologies or programming languages**: It is not uncommon that different parts of the system can be written in different programming languages or use fundamentally different technologies, libraries, and frameworks.

- **Independent APIs**: Components may provide fully independent APIs.

- **Code review process**: Some components may be subject to a stricter code review process and additional requirements.

- **Security**: Components may have different security requirements and may require additional isolation from the rest of the application for security reasons.

- **Compliance**: Some parts of the system may be subject to stricter compliance requirements. For example, handling **personally identifiable information** (**PII**) for users from a certain region can put stricter requirements on the entire system. Logical separation of such components helps to reduce the scope of work required to keep the system compliant.

With all the preceding issues described, we can see that at a certain point monolithic applications can become too big for a *one-size-fits-all* model. As the application grows, certain parts of it may start becoming independent and have different requirements, benefiting from a logical separation from the rest of the application.

In the next section, we are going to see how splitting the application into microservices can solve the aforementioned problems and which aspects of it you should be careful with.

Pros and cons of microservices

In order to understand how to get the best results from using microservices and which issues to be aware of, let's review the pros and cons of the microservice model.

Benefits of microservices

As previously described, different application components may have fundamentally different requirements and at certain points diverge so much that it would be beneficial to separate them. In this case, microservice architecture provides a clear solution by decoupling the parts of the system.

Microservices provide the following benefits to developers:

- **Faster compilation and build time**: Faster build and compilation time may play a key role in speeding up all development processes.

- **Faster deployments, lower deployable size**: When each part of the system is deployed separately, the deployable size can get so significantly smaller that individual deployments can take just a fraction of the time compared to monolithic applications.

- **Custom deployment cadence**: The microservice model solves the problem of following a custom deployment schedule. Each service can be deployed independently and follow its own schedule.

- **Custom deployment monitoring**: Some services can perform more critical roles in the system than others and may require more fine-grained monitoring and extra checks.

- **Independent and configurable automated testing**: Services may be configured to perform different automated tests as a part of the build and deployment pipeline. Additionally, the scope of checks can be reduced for individual microservices, that is, we don't need to perform tests for the entire application, which may take longer.

- **Cross-language support**: It is no longer required to run an application as a single executable, so it is possible to implement different parts of the system using different technologies, finding the best fit for each problem.

- **Simpler APIs**: Fine-grained APIs are one of the key aspects of microservice development and having clear and efficient APIs helps to enforce the right composition of the system.

- **Horizontal scaling**: Microservices are easier and often cheaper to scale horizontally. Monolithic applications are usually resource-heavy and running them on numerous instances could be quite expensive due to high hardware requirements. Microservices, however, can be scaled independently. So, if a particular part of the system requires running on hundreds or thousands of servers, other parts don't need to follow the same requirements.

- **Hardware flexibility**: Splitting an application often means reducing the hardware requirements for most parts of the system. It provides more flexibility in choosing the hardware or cloud providers to execute applications.

- **Fault isolation**: Service decoupling provides an efficient safety mechanism to prevent major issues on partial system failures.

- **Understandability**: Services are easier to understand and maintain due to lower code base sizes.

- **Cost optimization**: Running most application components on lower-grade instances compared to expensive high-resource monolithic instances may result in significant cost savings for the company.

- **Distributed development**: Removing the coupling between the components helps achieve more independence in code development, which can play an important role in distributed teams.

- **Ease of refactoring**: In general, it is much easier to perform refactoring for microservices due to the lower scope of changes and independent release and testing processes, which helps detect possible issues and reduce the scope of failures.

- **Technological freedom**: With microservice architecture, it is much easier to switch to new technologies given that each service is smaller in size and is structurally independent of the others. This can play a key role in companies with an open and experimental development culture, helping find the right solutions for particular problems and keep their technological stack up to date.

- **Independent decision-making**: Developers are free to choose programming languages, libraries, and tools that fit their needs the best. This does not, however, imply that there should be no standardization, but it is often highly beneficial to achieve a certain degree of freedom for distributed decision-making.

- **Removing unnecessary dependencies**: It is easy to miss detecting unwanted dependencies between the components of a monolithic application given the tighter coupling of the components. Microservice architecture helps you notice unwanted dependencies between components and restricts the use of certain services to particular parts of the application.

As we can see, microservices bring a high degree of flexibility and help to achieve a higher level of independence between the components. These aspects may be instrumental to the success of a large development team, allowing them to build and maintain independent components separately. However, any model comes at its own cost, and in the next section, we are going to see the challenges you could face with a collection of microservices.

Common issues of microservices

As with any solution, microservice architecture has its own issues and limitations. Some issues with microservice architecture include the following:

- **Higher resource overhead**: When an application consists of multiple components, instead of sharing the same process space, there is a need to communicate between the components that involve higher network use. This puts more load on the entire system and increases traffic, latency, and I/O usage. In addition, the total CPU and RAM are also higher due to the extra overhead of running each component separately.

- **Debugging difficulty**: Troubleshooting and debugging are often more difficult when you deal with multiple services. For example, if multiple services process a request that fails, a developer needs to access the logs of multiple services in order to understand what caused the failure.

- **Integration testing**: Separating a system requires building a large set of integration tests and other automated checks that would monitor the compatibility and availability of each component.

- **Consistency and transactions**: In microservice applications, the data is often scattered across the system. While this helps to separate the independent parts of the application, it makes it harder to do transactional and atomic changes in the system.

- **Divergence**: Different services may use different versions of libraries, which may include incompatible or outdated ones. Divergence makes it harder to perform system upgrades and resolve various issues, including software vulnerability fixes.

- **Tech debt addressability**: It is much harder to address tech debt in a distributed system where each component is owned by a different team.

- **Observability**: Managing multiple applications brings additional challenges in collecting and using the system events and messages, including logs, traces, and metrics. Developers need to make sure all such signals are collected for all applications and are available for analysis, including all necessary contextual information to debug any issues and locate the root cause of the issue among the target services.

- **Possible duplication, overlapping functionality**: In a highly distributed development environment, it is not uncommon to have multiple components performing similar roles in the system. It is important to set clear boundaries within the system and decide in advance which particular roles the components are assigned.

- **Ownership and accountability**: Ownership becomes a major aspect of the development process when there are many different teams maintaining and developing independent components. It is crucial to define clear ownership contracts to address the development requests, security and support issues, and all other types of maintenance work.

As we have just illustrated, the microservice model comes at a cost and you should expect that you will need to solve all these challenges at a certain point. Being aware of the possible challenges and being proactive in solving them is the key to success – the benefits that we have described earlier can easily outweigh the possible issues.

In the next section, we are going to summarize when to use the microservices and learn some best practices for working with them.

When to use microservice architecture

We have covered the benefits and common issues of microservices, providing a good overview of the applicability of using the microservice architecture model in the application. Let's summarize the key points of using the microservice model, which are the following:

- **Don't introduce microservices too early**: Don't use the microservice architecture too early if the product is loosely defined or can go through significant changes. Even when developers know the exact purpose of the system, there are high chances of various changes in the early stages of the development process. Starting from a monolithic application – and splitting it over time once there are clearly defined business capabilities and boundaries – helps reduce the amount of work and establish the right interfaces between the components.

- **No size fits all**: Each company is unique and the final decision should depend on many factors, including the size of the team, its distribution, and geography. A small local team may be comfortable working with a monolithic application, whereas a geographically distributed team may highly benefit from splitting the application into multiple microservices to achieve higher flexibility.

Additionally, let's summarize the best practices of using the microservice architecture model for applications, which are the following:

- **Design for failure**: In a microservice architecture, there are many interactions between the components, most of which are happening via remote calls and events. This increases the chance of various failures, including network timeouts, client errors, and many more. Build the system thinking of every possible failure scenario and different ways to proceed with it.

- **Embrace automation**: Having more independent components requires much stricter checks in order to achieve stable integration between the services. Investing in solid automation is absolutely necessary in order to achieve a high degree of reliability and ensure all changes are safe to be deployed.

- **Don't ship hierarchy**: It is a relatively common practice to split the application into services based on the organizational structure, where each team may be responsible for its own service. This model works well if the organizational structure perfectly aligns with the business capabilities of the microservices, but quite often this is not the case. Instead of using a service-per-team model, try to define the clear domains and business capabilities around which the code is structured and see how the components interact with each other. It is not easy to achieve perfect composition, but you will be highly rewarded for it.

- **Invest in integration testing**: Make sure you have comprehensive tests for the integrations between your microservices performing automatically.

- **Keep backward compatibility in mind**: Always remember to keep your changes backward compatible to ensure that new changes are safe to deploy. Additionally, use techniques such as versioning, which we are going to cover in the next chapter of the book.

At this point, we have covered the key aspects of microservice development, and you have learned its benefits and the challenges you may face. Before we proceed to the next chapter, let's cover one more topic to ensure we are ready for the journey into microservice development. Let's get familiar with the Go programming language and its role in microservice development.

Role of Go in microservice development

Over the last decade, the **Go** programming language has become one of the most popular languages for application development. There have been many factors contributing to its success, including its simplicity, ease of writing network applications, and an ability to easily develop parallel and concurrent applications.

Additionally, the larger developer community has played a key role in raising its popularity across all types of developers. The Go community is welcoming to everybody, from people just starting their journeys into programming to seasoned experts with decades of experience building different types of applications.

The Go standard library provides a set of packages that can often be enough for building a complete web application or an entire service, sometimes without even requiring any external dependencies. Many developers have been fascinated by the ease of writing applications and tools performing network calls, data serialization and encoding, file processing, and many other types of common operations.

This simplicity, paired with fast and efficient compilation into native binaries as well as rich tooling, made it one of the primary languages for writing web tools and services. The high adoption of the Go language for web service development made it one of the primary choices for writing microservices across the industry.

The biggest advantages of Go for microservice development include the following:

- **Smooth learning curve**: As one of the critical aspects of application development in growing teams, the simplicity of the Go language helps reduce the onboarding time for new, inexperienced developers.

- **Explicit error handling**: While being a hot topic in the Go community, error handling in Go encourages explicit handling of all application errors. It aligns with one of the key principles of microservice development of designing applications for failure.

- **Useful standard library**: The Go standard library includes lots of packages that can be used in production-grade systems without requiring external solutions.

- **Community support**: The Go community is among the biggest in the industry and the most popular libraries get enough support and maintenance.

- **Ease of writing concurrent code**: Concurrent calls are very common in microservice application logic - microservices often call multiple other services and combine their results. Writing concurrent code in **Golang** can be a fairly trivial task when utilizing the built-in sync package and core language features, such as channels and Goroutines.

The growth of the Go community has increased the rate of development of additional libraries for the language and resulted in the creation of the entire ecosystem of tools, powering application logging, debugging, and implementations of all widely used networking protocols and standards. The community keeps growing and the rate of new releases is only accelerating.

Summary

We have discussed the key aspects of the microservice development model, including the motivation to use it, the common benefits, and the possible challenges. You have learned that the microservice model brings many advantages, helping you achieve a higher degree of flexibility. It also comes with its own costs, which often include additional complexity and a lack of uniformness in the system. Microservice architecture requires you to think about these problems proactively in order to address them before they become big issues.

In the next chapter, we are going to start our journey into microservice development with the Go language. You will learn the important basics of the Go programming language and we will scaffold our microservices, which we are going to improve throughout the rest of the book.

Further reading

- **A collection of resources on microservice development**: `https://microservices.io/`

- **Overview of the microservice architecture model**: `https://martinfowler.com/articles/microservices.html`

- **15 best practices for building microservices**: `https://www.bmc.com/blogs/microservices-best-practices/`

Part 2: Foundation

This part covers the foundational aspects of Go microservice development, such as service discovery, data serialization, synchronous and asynchronous communication, deployment, and testing. You will learn how to scaffold Go microservices, establish communication between them, store service data, and implement service APIs, as well as many other important aspects of microservice development.

This part contains the following chapters:

- *Chapter 2, Scaffolding a Go Microservice*
- *Chapter 3, Service Discovery*
- *Chapter 4,* Serialization
- *Chapter 5,* Synchronous Communication
- *Chapter 6,* Asynchronous Communication
- *Chapter 7,* Storing Service Data
- *Chapter 8,* Deployment with Kubernetes
- *Chapter 9,* Unit and Integration Testing

2
Scaffolding a Go Microservice

In this chapter, we will finally start scaffolding our microservice code. The goal of this chapter is to establish a solid foundation for writing Go microservices and setting the right structure for future changes. While Go makes it relatively easy to write small applications, there are multiple challenges that engineers may face along the way, including the following:

- How to set the right project structure to make it easier to evolve and maintain the code base
- How to write idiomatic Go code that is going to be consistent with the largest Go code bases
- How to separate the components of a microservice and wire them together

In this chapter, we are going to address each of these challenges. First, you will be introduced to the key aspects of writing idiomatic and conventional Go code. You will learn important recommendations for writing and organizing your code base, as well as how to set up the proper code structure for your services. Then, we are going to introduce you to an example application, which will consist of three microservices that we are going to use throughout the book. In the following chapters, we will add additional features to these services, illustrating all the important areas of microservice development.

In this chapter, we will cover the following topics:

- Go basics
- Project structure
- Scaffolding an example application

Technical requirements

To complete this chapter, you need to have Go 1.11 or above. If you don't have Go installed, you can download it from the official website at `go.dev/dl`.

You can find the code examples for this chapter on GitHub: `https://github.com/PacktPublishing/microservices-with-go/tree/main/Chapter02`.

Go basics

Go is a great language for writing microservices. It is relatively easy to learn and has a pretty smooth learning curve, making onboarding new engineers easier. While you may have already had some experience with Go, one of the purposes of this book is to provide enough information to all types of developers—from beginners to highly experienced professionals.

In this section, we are going to summarize important concepts of the language. If you already have experience with Go, you can still quickly scan through this part. It also includes some useful recommendations and best practices commonly missed even by experienced engineers.

Core principles

Before we proceed to look at the basics of Go, I'm going to share with you some fundamental principles that will help you make decisions when writing and organizing your code. These principles include the following:

- *Always follow the official guidelines.* It is not uncommon for us engineers to have strong opinions about various styling and coding practices. However, in any developer community, consistency is more important than individual opinions. Make sure you get familiar with the most fundamental Go programming guidelines, written by the Go team:

  - **Effective Go—an official set of guidelines for Go developers**: `https://go.dev/doc/effective_go`

  - **Go code review comments**: Another useful source of information on Go development, covering various aspects, including code style, naming, and error handling: `https://github.com/golang/go/wiki/CodeReviewComments`

- *Follow the style used in the standard library.* The standard Go library, which comes with any Go installation, is the best source of code examples and comments. Get familiar with some of the packages from the library, such as `context` and `net`. Following the coding style used in these packages will help you to write consistent, readable, and maintainable code, regardless of who will be using it later.

- *Do not try to apply the ideas from other languages to Go.* Instead, understand the philosophy of Go and see the implementation of the most elegant Go packages—you can check the `net` package for some good examples: `https://pkg.go.dev/net`.

Now, as we are aligned on the core principles, let's move on to the key recommendations for writing conventional and idiomatic Go code.

Writing idiomatic Go code

This section summarizes the key topics described in the *Effective Go* document. Following the suggestions provided in this section will help you to keep your code consistent with the official guidelines.

Naming

Naming is one of the most important aspects of Go development. Writing Go code in an idiomatic way requires an understanding of its core naming principles:

- Exported names start with an uppercase character.

- When a variable, struct, or interface is imported from another package, its name includes a package name or alias, for example, `bytes.Buffer`.

- Since references include package names, you should not prefix your names with the package name. If the package name is `xml`, use the name `Reader`, not `XMLReader`—in the second case, the full name would be `xml.XMLReader`.

- Packages are generally given lowercase, single-word names.

- It is not idiomatic to start the names of getters with the `Get` prefix. If your function returns the user's age, call the function `Age()`, not `GetAge()`. Using the `Set` prefix, however, is fine; you can safely call your function `SetAge()`.

- Single-method interfaces are named using the method name plus an `er` suffix. For example, an interface with a `Write` function would be called `Writer`.

- Initialisms and acronyms should have a consistent case. The correct versions would be `URL`, `url` and `ID`, include while `Url`, `Id` would be incorrect.

- Variable names should be short rather than long. In general, follow this simple rule—the closer to declaration a name is used, the shorter it should be. For iterating over an array, use *i* for the index variable.

Additional naming recommendations include the following:

- The package name should be short, concise, and evocative and should provide context for its contents, for example, `json`.

- Keep the contents of a package consistent with the name. If you start noticing that a package includes extra logic that has no relationship to the package name, consider exporting it to a separate one or using a more descriptive name.

- Use name abbreviations only if they are widely used (for example, `fmt` or `cmd`).

- Avoid name collisions, when possible. For example, if you introduce a set of string functions, avoid calling it `strings` package because a package with the same name exists in the Go standard library and is already widely used.

- Consider the client's point of view when giving names to your code. Think about how the code is going to be used when giving a name to it, for example, the `Writer` interface for proving the write functionality.

In addition to these rules, remember to keep the naming consistent across your code base. It will help make it easier to read and write new code—good names will act as examples for other engineers as well.

Comments

Comments are the next important aspect of Go development. Go comments can be used in two different ways:

- Seeing the comments alongside the code
- Viewing the package documentation generated by the `godoc` tool

General principles for Go comments include the following:

- Every package should have a comment describing its contents.
- Every exported name in Go should have a comment.
- Comments should be complete sentences and end with a period.
- The first sentence of the comment should start with the name being exported and provide a summary of it, as in the following example:

```
// ErrNotFound is returned when the record is not found.
var ErrNotFound = errors.New("not found")
```

The Go standard library provides many good examples of code comments, so I always suggest getting familiar with some examples from it.

Errors

General recommendations for Go errors include the following:

- Only use panics in truly exceptional cases.
- Always handle each error; don't discard errors by using _ assignment.
- Error strings should start with a lowercase character, unless they begin with names requiring capitalization, such as acronyms.
- Error strings, unlike comments, should not end with punctuation marks, as in the following example:

```
return errors.New("user not found")
var errUserNotFound = errors.New("user not found")
```

- When calling a function returning an error, always handle the error first.

- Wrap errors if you want to add additional information to the clause. The conventional way of wrapping errors in Go is to use %w at the end of the formatted error:

```
if err != nil {
    return fmt.Errorf("upload failed: %w", err)
}
```

- While checking for errors, using the == operator may result in improper handling of the wrapped errors. There are two solutions to this. For a comparison to a sentinel error, such as errors.New("some error"), use errors.Is:

```
if errors.Is(err, ErrNotFound) {
    // err or some error it wraps is ErrNotFound.
}
```

For error types, use errors.As:

```
var e *QueryError
if errors.As(err, &e) {
    // err has *QueryError type.
}
```

Additionally, keep errors descriptive yet compact. It should be always easy to understand what exactly went wrong by reading the error message.

Interfaces

Key principles of Go **interfaces** include the following:

- Do not define interfaces before they are used without a realistic example of usage.

- Return concrete (using a pointer or struct) types instead of an interface in your functions.

- Single-method interfaces should be called by the method name and include the er suffix, for example, the Writer interface with a Write function.

See some built-in interfaces, such as Writer and Reader, to get a good example of defining and using interfaces in Go.

Tests

We are going to cover testing in detail in *Chapter 8* of this book. Let's provide here some key suggestions for writing Go tests in an idiomatic way:

- Tests should always provide information to the user on what exactly went wrong in case of a failure.
- Consider writing table-driven tests whenever possible. See this example: `https://github.com/golang/go/blob/master/src/fmt/errors_test.go`.
- Generally, we should only test public functions. Your private function should be indirectly tested through them.

Make sure you always write tests for your code. Not only does this help with finding bugs earlier but it also helps to see how your code can be used. I personally find the latter especially useful.

Context

One of the key differences between the Go language and other popular languages is explicit context propagation. **Context propagation** is a mechanism of propagating an additional call argument, called **context**, into function calls, passing additional metadata.

Go context has a type called `context.Context`. There are multiple ways of using it:

- **Cancelation logic**: You can pass a special instance of a context that can get *canceled*. In that case, all functions you were to call with it would be able to detect this. Such logic can be useful for handling application shutdown or stopping any processing.
- **Timeouts**: You can set the timeouts for your execution by using the corresponding context functions.
- **Propagating extra metadata**: You can propagate additional key-value metadata inside the context. This way, any downstream functions called would receive that metadata inside the context object. There are some useful applications of this approach, one of which is distributed tracing, which we are going to cover in the following chapters.

We will get back to context propagation in the following chapters. Now, we can define some important aspects of using context in Go:

- Context is immutable but can be cloned with extra metadata.
- Functions using context should accept it as their first argument.

Additionally, some context best practices are as follows:

- Always pass context to functions performing I/O calls.

- Limit the usage of context for passing any metadata. You should use metadata propagation for truly exceptional cases, such as distributed tracing, mentioned earlier.

- Do not attach context to structures.

Now, as we have discussed the key recommendations for writing idiomatic Go code, we can move on to the next section, which is going to cover the project structure recommendations and standards for Go applications.

Project structure

The project structure is the foundation of and plays a major role in the readability and maintainability of your code. As we discussed in the previous sections, in Go projects, the structure may play a more important role than in other languages, because each exported name generally includes the name of its package. This requires you to have good and descriptive naming for your packages and directories, as well as the right hierarchy of your code.

While the official guidelines define some strong recommendations for naming and coding styles there aren't that many rules constraining the Go project structure. Each project is unique by its nature, and developers are generally free to choose the way they organize the code. However, there are some common practices and specifics of Go package organization that we are going to cover in this section.

Private packages

In Go, all code stored inside a directory called `internal` can be imported and used only by packages stored within the same directory or one of the directories it includes. Putting code into an internal directory can ensure your code is not exported and used by external packages. This can be useful for the following different cases:

- Hide the details of the implementation from the user if some of the types of functions need to be exported.

- Ensure no external package relies on your types and functions, which you don't want to expose widely.

- Remove possible unnecessary dependencies between the packages.

- Avoid extra refactoring and maintenance difficulties if your code is unexpectedly used by other developers/teams.

I have found it useful to use internal packages as a protection against unwanted dependencies. This plays a big role in large repositories and applications, where there is a high possibility of unexpected dependencies between the packages. Large code bases that don't have a separation between private and public packages often suffer from an effect called *spaghettification*—when packages depend on each other in an uncontrolled and chaotic way.

Public packages

There is another type of directory name with a semantic meaning in Go—a directory called pkg. It implies that it is OK to use the code from this package externally.

The pkg directory isn't recommended officially, but it is widely used. Ironically, the Go team used this in the library code and then got rid of this pattern, while the rest of the Go community adopted it so widely that it became a common practice.

It is up to you whether you use a pkg directory in your applications. But in tandem with the internal directory, it can help to organize your code so that what is private and what is public is clear, easing the code navigation for the developers.

Executable packages

The cmd package is commonly used in the Go community to store the code of one or multiple executable packages with a main function. This may include the code starting your application or any code for your executable tools. For a single-app directory, you can store your Go code directly in the cmd package:

```
cmd/
cmd/main.go
```

For a multi-app directory, you can include subpackages in cmd packages:

```
cmd/
cmd/indexer/main.go
cmd/crawler/main.go
```

Other commonly used directories

The following list includes some other commonly used directory or package names in the Go community:

- api: JSON schema files and definitions in various protocols, including gRPC. We are going to cover these topics in *Chapter 4*.
- testdata: Files containing the data used in tests.
- web: Web application components and assets.

Common files

Here is a list of common filenames, which will keep your packages consistent with the official library and lots of third-party libraries:

- main.go: A file containing the main() function

- `doc.go`: Package documentation (a separate file is not necessary for small packages)
- `*_test.go`: Test files
- `README.md`: A read-me file written in the Markdown language
- `LICENSE`: A license file, if there is one
- `CONTRIBUTING.md`/CONTRIBUTORS/AUTHORS: List of contributors and/or authors

Now, let's cover the best practices for organizing the code base for Go applications.

Best practices

In this section, you can find a list of best practices for organizing the Go application project structure. It is going to help you to keep your code aligned with thousands of other Go packages and keep it conventional and idiomatic. The best practices of Go project organization include the following:

- Separate private code using an internal directory.
- Get familiar with the way popular open source Go projects, such as `https://github.com/kubernetes/kubernetes`, are organized. This can provide you with great examples of how to structure your repository.
- Split in a sufficiently granular way. Don't split the packages too early but also avoid having a lot of logic in a single package. Generally, you will find that the easier it is to give a short and specific self-descriptive name to a package, the better your code composition is.
- Avoid long package names.
- Always be ready to change the structure if requirements are changed or if the structure no longer reflects the package name/original intent.

This sums up the part of the chapter describing the core principles and best practices of Go application development and code organization. Now, we are ready to get to the practical side of this chapter.

Scaffolding an example application

We have covered the general recommendations for writing and organizing Go applications and we are finally ready to start writing the code! In this section, we are going to introduce an application, consisting of multiple microservices that are going to be used throughout the book. In each chapter, we are going to add to or improve them, converting them from small examples into production-grade services that are ready to be used.

You will learn how to scaffold microservice code and split the code into separate logical parts, each having its own role. We are going to apply the project structure and Go knowledge you gained in this chapter to illustrate how to set the right structure for each service and write its code in a conventional and idiomatic way.

Movie application

Let's imagine we are building an application for movie lovers. The application would provide the following features:

- Get the movie metadata (such as title, year, description, and director) and the aggregated movie rating

- Rate a movie

All the listed features seem to be closely related. However, let's take a closer look at them.

Movie metadata

Let's assume we have the metadata for a collection of movies, which includes the following fields:

- ID
- Title
- Year
- Description
- Director
- List of actors

Such information about movies doesn't generally change unless somebody wants to update the description, but for simplicity, we may assume that we are dealing with a static dataset. We would retrieve the records based on their IDs, so we could use any key-value or document database to store and access the metadata.

Ratings

Let's now review the functionality required for storing and retrieving movie ratings.

Generally, we would need to perform the following rating operations:

- Store a movie rating
- Get the aggregated movie rating

Later, we would also need to support rating deletion, but for now, we can just keep this logic in mind while designing the application.

The ratings data is quite different from the movie metadata—we can both append and delete the records. In addition to this, we need to return the aggregated rating, so we should either be able to return all stored ratings for an item and perform the aggregation on the go or have separate logic for performing and storing the aggregations. You will notice that the ways we access ratings and movie metadata are different. This hints that the ratings data can, and probably should, be stored separately from the movie metadata.

While designing the application, it is beneficial to think one step ahead and imagine how the application may evolve in the future. This does not mean that you should necessarily build the application trying to predict future use cases, because it can lead to unnecessary abstractions that may not be needed later if your plans change. However, thinking one step ahead may save you time later if you find more efficient ways of modeling and storing your data, which would help you to adapt to changing requirements.

Let's see how the rating service can possibly evolve. At some point, we may want to extend the rating functionality to other types of movie-related records. A user may be able to do the following:

- Rate the actor's performance in some movies
- Rate the movie soundtrack
- Rate the movie's costume design

When making decisions on supporting future use cases, you should ask yourself, how likely it is that I will need to implement that logic in the observable future (6 to 12 months)? You should generally avoid thinking much further ahead because the requirements and goals may change. However, if you are quite certain you have plans to support particular features, you should make sure your data model can support those features without major changes.

Let's assume we definitely want to implement the additional ratings mentioned earlier. In this case, we want to make sure we can design our application in a way that would support the ratings for different types of objects.

Let's define the API for such a rating component:

- Store the rating record, including the following:

  - ID of the user who gave the rating
  - Type of record
  - ID of the record
  - Rating value

- Get the aggregated rating for a record by its ID and type.

This API supports record types, so we can easily add more types of ratings without changing the system. The trade-off we made here is quite reasonable—the API is different on the basis of just one field (record type) from the API of a rating system designed just for movies. However, this gives us complete freedom in introducing new rating types in the future! Such a trade-off seems very reasonable given that we have decided we will certainly need those ratings in the future.

Should we split the application?

Let's provide a summary of the two parts of the application we have just described:

- Movie metadata:

  - Retrieve the metadata for a movie by its ID.

- Ratings:

  - Store a rating for a record.

  - Retrieve an aggregated rating for a record.

After we abstracted the rating component by letting it support various record types, it stopped being a movie rating component and became a more generic record rating system. The movie metadata component is now loosely coupled to the rating system—the rating system can store the ratings for movies as well as for any other possible types of records.

As we discussed previously, the data models for both components are also quite different. The movie metadata component stores static data, which is going to be retrieved by ID, while the rating component stores dynamic data, which requires aggregation.

Both components seem to be relatively independent of each other. This is a perfect example of a situation where we may benefit from splitting the application into separate services:

- Logic is loosely coupled

- Data models are different

- Data is generally independent

This list is not complete, and you need to consider all the aspects described in *Chapter 1*, to make a decision on splitting the application. However, since this book covers microservice development, let's make our decision here and decide to split the system into separate services.

Let's list the services we would split the application into:

- **Movie metadata service**: Store and retrieve the movie metadata records by movie IDs.

- **Rating service**: Store ratings for different types of records and retrieve aggregated ratings for records.

- **Movie service**: Provide complete information to the callers about a movie or a set of movies, including the movie metadata and its rating.

Why did we end up with three services here? We did this for the following reasons:

- The movie metadata service would be solely responsible for accessing the movie metadata records.

- The movie service would provide the client-facing API, aggregating two separate types of records—movie metadata and ratings. The records would be stored in two separate systems, so this component would join them together and return to the caller.

- If we introduce any other types of records in the system, such as likes, reviews, and recommendations, we will plug them into the movie service, not the movie metadata service. The movie metadata service would be used solely for accessing the static movie metadata, not any other types of records.

- Movie metadata services can potentially evolve in the future by getting more metadata-related functionality, such as editing or adding descriptions in different languages. This also hints that it is better to keep this component solely for the metadata-related features.

Let's illustrate these services in a diagram:

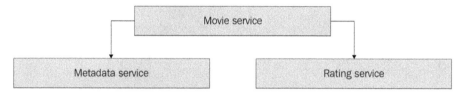

Figure 2.1 – Movie application services

Now, as we have definitions of the three microservices, let's finally proceed to the coding part.

Application code structure

Let's align on how are we going to structure the code of all microservices in relation to each other. I would suggest storing them inside a single directory, which would be our application root. Create a new directory (you may call it movieapp), and inside it, create the following directories for our microservices:

- rating
- metadata
- movie

Throughout the book, I will use the directory paths relative to the application directory you've created, so when you see a directory or filename, assume it is stored in the app directory you chose for this.

From the *Project structure* section, we know that the logic containing the main function generally resides in the cmd directory. We will use this approach in our microservices—for example, the main file for the rating service would be called rating/cmd/main.go.

Each service may contain one or multiple packages related to the following logical roles:

- API handlers
- Business/application logic
- Database logic
- Interaction with other services

Note that handler and business/application logic are separate, even though the primary purpose of the application may be to handle the API requests. This is not absolutely necessary, but it's a relatively good practice to separate the business logic from the API handling layer. This way, if you migrate from one type of API to another (for example, from HTTP to gRPC), or support both, you don't need to implement the same logic twice or rewrite it. Instead, you would just call the business logic from your handler, keeping the handler as simple as possible and making its primary purpose to pass the requests to the relevant interfaces.

We can illustrate this relationship with the help of a diagram:

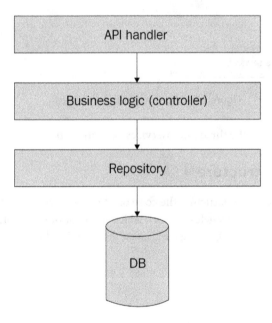

Figure 2.2 – Layers of a service

As you can see in the diagram, the API handler does not access the database directly. Instead, the database access is performed on a business logic layer.

There is no convention in the Go community on how to call packages serving these purposes, so we are free to choose the names for our packages providing such logic. It is, however, important that you keep these names consistent across all your microservices, so let's align on a common naming convention for these types of packages.

In this book, we are going to use the following names for our application components:

- **controller**: Business logic

- **gateway**: Logic for interacting with other services

- **handler**: API handlers

- **repository**: Database logic

Now, since we are aligned on the naming, let's proceed to the last step of setting up our project. Execute this command in the application root directory:

```
go mod init movieexample.com
```

This command creates a Go module called `movieexample.com`. A Go module is a collection of related packages stored in a file tree. They help manage dependencies for your project, and we are going to use this feature in all the chapters.

Now, we can proceed to code scaffolding for our first microservice.

Movie metadata service

Let's summarize the logic of the movie metadata service:

- **API**: Get metadata for a movie

- **Database**: Movie metadata database

- **Interacts with services**: None

- **Data model type**: Movie metadata

This logic would translate into the following packages:

- `cmd`: Contains the main function for starting the service

- `controller`: Our service logic (read the movie metadata)

- `handler`: API handler for a service

- `repository`: Logic for accessing the movie metadata database

Let's store the logic of our service in a directory called `metadata`. Following the conventions we described earlier in the chapter, the executable code containing the main file is going to be stored in the `cmd` package. All code that we are not going to export will be stored in the `internal` directory and this will include most of our applications. The exported structures will reside in the `pkg` directory.

Applying the rules that we just described, we are going to structure our packages in the following way:

- `metadata/cmd`
- `metadata/internal/controller`
- `metadata/internal/handler`
- `metadata/internal/repository`
- `metadata/pkg`

Once you have created the directories listed here, let's proceed to implement the code for our microservice.

Model

First, we are going to implement the structure for the movie metadata. Inside the `metadata/pkg` directory, create a `metadata.go` file using the following code:

```
package model

// Metadata defines the movie metadata.
type Metadata struct {
    ID          string `json:"id"`
    Title       string `json:"title"`
    Description string `json:"description"`
    Director    string `json:"director"`
}
```

This structure is going to be used by the callers of our service. It includes JSON annotations, which we are going to use later in this chapter.

Repository

Now, let's create the stub logic for handling the database logic. Inside the `metadata/internal/repository` directory, add an `error.go` file using the following code:

```
package repository

import "errors"

// ErrNotFound is returned when a requested record is not
// found.
var ErrNotFound = errors.New("not found")
```

This file defines an error for the case when the record is not found. We are going to use this error in our implementation.

In the next step, we are going to add the repository implementation. Even if you have some specific technology to work with for storing the data, it is often useful to provide more than one implementation of the database logic. I always find it useful to include an in-memory implementation of the database logic that can be used for testing and local development, reducing the need for any additional databases or extra libraries. I am going to illustrate how to do this.

Inside the `metadata/internal/repository` directory, create a directory called `memory` that will contain the in-memory implementation or our movie metadata database. Add a `memory.go` file to it, using the following code:

```
package memory

import (
    "context"
    "sync"

    "movieexample.com/metadata/internal/repository"
    "movieexample.com/metadata/pkg/model"
)

// Repository defines a memory movie metadata repository.
type Repository struct {
    sync.RWMutex
    data map[string]*model.Metadata
}

// New creates a new memory repository.
func New() *Repository {
    return &Repository{data: map[string]*model.Metadata{}}
}

// Get retrieves movie metadata for by movie id.
func (r *Repository) Get(_ context.Context, id string) (*model.Metadata, error) {
    r.RLock()
    defer r.RUnlock()
```

```
    m, ok := r.data[id]
    if !ok {
        return nil, repository.ErrNotFound
    }
    return m, nil
}


// Put adds movie metadata for a given movie id.
func (r *Repository) Put(_ context.Context, id string, metadata
*model.Metadata) error {
    r.Lock()
    defer r.Unlock()
    r.data[id] = metadata
    return nil
}
```

Let's highlight some aspects of the code we've just added:

- First, we called the Repository structure because it provides a good name to the users when combined with the name of its package— memory.Repository.

- Second, we used the exported ErrNotFound that we previously defined, so callers can check their code. It is usually good practice to do so because it allows the developers to check for a specific error in their code. We will illustrate how to write tests for it in *Chapter 8*.

- Additionally, the function creating the repository is called New. This is often a good name for short packages when there is just one type being created.

- Our Get and Put functions accept context as the first argument. We mentioned this approach in the *Writing idiomatic Go code* section—all functions performing I/O operations must accept context.

- Our implementation is using a sync.RWMutex structure to protect against concurrent writes and reads

Now, let's move on to the business logic layer.

Controller

The next step is to add a controller to encapsulate our business logic. Even if your logic is trivial, it is still a good practice to keep it separate from the handler from the beginning. This will help you avoid further changes and, more importantly, keep the structure of your applications consistent.

Inside the metadata/internal/controller package, add a directory called metadata. Inside it, add a controller.go file with the following logic:

```go
package metadata

import (
    "context"
    "errors"

    "movieexample.com/metadata/internal/repository"
    "movieexample.com/metadata/pkg/model"
)

// ErrNotFound is returned when a requested record is not
// found.
var ErrNotFound = errors.New("not found")

type metadataRepository interface {
    Get(ctx context.Context, id string) (*model.Metadata,
error)
}

// Controller defines a metadata service controller.
type Controller struct {
    repo metadataRepository
}

// New creates a metadata service controller.
func New(repo metadataRepository) *Controller {
    return &Controller{repo}
}

// Get returns movie metadata by id.
func (c *Controller) Get(ctx context.Context, id string)
(*model.Metadata, error) {
    res, err := c.repo.Get(ctx, id)
    if err != nil && errors.Is(err, repository.ErrNotFound) {
```

```
        return nil, ErrNotFound
    }
    return res, err
}
```

The controller we created is currently just a wrapper around the repository. However, the controller will generally have more logic, so it is preferable to keep it separate.

Handler

Now, we are going to create the API handler. Inside the metadata/internal/handler directory, create a directory called http. Inside it, create a file called http.go with the following logic:

```go
package http

import (
    "encoding/json"
    "errors"
    "log"
    "net/http"

    "movieexample.com/metadata/internal/controller/metadata"
    "movieexample.com/metadata/internal/repository"
)


// Handler defines a movie metadata HTTP handler.
type Handler struct {
    ctrl *metadata.Controller
}

// New creates a new movie metadata HTTP handler.
func New(ctrl *metadata.Controller) *Handler {
    return &Handler{ctrl}
}
```

Now, let's implement the logic for retrieving movie metadata:

```go
// GetMetadata handles GET /metadata requests.
func (h *Handler) GetMetadata(w http.ResponseWriter, req *http.
```

```
Request) {
    id := req.FormValue("id")
    if id == "" {
        w.WriteHeader(http.StatusBadRequest)
        return
    }
    ctx := req.Context()
    m, err := h.ctrl.Get(ctx, id)
    if err != nil && errors.Is(err, repository.ErrNotFound) {
        w.WriteHeader(http.StatusNotFound)
        return
    } else if err != nil {
        log.Printf("Repository get error: %v\n", err)
        w.WriteHeader(http.StatusInternalServerError)
        return
    }
    if err := json.NewEncoder(w).Encode(m); err != nil {
        log.Printf("Response encode error: %v\n", err)
    }
}
```

The handler we just created uses our repository to retrieve the information and return it in JSON format. We chose JSON here just for simplicity. In *Chapter 4*, we are going to cover more data formats and illustrate how they can benefit your applications.

You may notice that we have called the package for our HTTP handler `http`. There is a trade-off here—while we are certainly colliding with its namesake standard library package, we get a pretty descriptive `http.Handler` exported name. Since our package is going to be used internally, this trade-off is reasonable.

Main file

Now, since we have created both a database and an API handler, let's create the executable for the metadata service. Inside the `metadata/cmd` directory, create the `main.go` file and add the following code:

```
package main

import (
    "log"
    "net/http"
```

```
        "movieexample.com/metadata/internal/controller/metadata"
        httphandler "movieexample.com/metadata/internal/handler/
http"
        "movieexample.com/metadata/internal/repository/memory"
)

func main() {
    log.Println("Starting the movie metadata service")
    repo := memory.New()
    ctrl := metadata.New(repo)
    h := httphandler.New(ctrl)
    http.Handle("/metadata", http.HandlerFunc(h.GetMetadata))
    if err := http.ListenAndServe(":8081", nil); err != nil {
        panic(err)
    }
}
```

The function we just created initializes all structures of our service and starts the http API handler we implemented earlier. The service is ready to process user requests, so let's move on to the other services.

Rating service

Let's summarize the logic of the rating service:

- **API**: Get the aggregated rating for a record and write a rating.
- **Database**: Rating database.
- **Interacts with services**: None.
- **Data model type**: Rating.

This logic would translate into the following packages:

- cmd: Contains the main function for starting the service
- controller: Our service logic (read and write ratings)
- handler: API handler for a service
- repository: Logic for accessing the movie metadata database

We are going to use exactly the same directory structure as we used for the metadata service:

- rating/cmd
- rating/internal/controller
- rating/internal/handler
- rating/internal/repository
- rating/pkg

Once you have created these directories, let's move on to the implementation of the service.

Model

Create a model directory inside rating/pkg and create a rating.go file, using the following code:

```go
package model

// RecordID defines a record id. Together with RecordType
// identifies unique records across all types.
type RecordID string

// RecordType defines a record type. Together with RecordID
// identifies unique records across all types.
type RecordType string

// Existing record types.
const (
    RecordTypeMovie = RecordType("movie")
)

// UserID defines a user id.
type UserID string

// RatingValue defines a value of a rating record.
type RatingValue int

// Rating defines an individual rating created by a user
for  // some record.
type Rating struct {
```

```
    RecordID     string        `json:"recordId"`
    RecordType   string        `json:"recordType"`
    UserID       UserID        `json:"userId"`
    Value        RatingValue   `json:"value"`
}
```

The file contains the model for our rating service, which is also going to be used by other services interacting with it. Note that we created separate types, RecordID, RecordType, and UserID. This will help us with readability and add extra type protection, as you will see in the implementation.

Repository

Create the in-memory implementation for our rating repository inside the rating/internal/repository/memory/memory.go file:

```go
package memory

import (
    "context"

    "movieexample.com/rating/internal/repository"
    "movieexample.com/rating/pkg/model"
)

// Repository defines a rating repository.
type Repository struct {
    data map[model.RecordType]map[model.RecordID][]model.Rating
}

// New creates a new memory repository.
func New() *Repository {
    return &Repository{map[model.RecordType]map[model.RecordID]
[]model.Rating{}}
}
```

Then, add an implementation of the Get function to it, as shown in the following code block:

```go
// Get retrieves all ratings for a given record.
func (r *Repository) Get(ctx context.Context, recordID model.
```

```
RecordID, recordType model.RecordType) ([]model.Rating, error)
{
    if _, ok := r.data[recordType]; !ok {
        return nil, repository.ErrNotFound
    }
    if ratings, ok := r.data[recordType][recordID]; !ok ||
len(ratings) == 0 {
        return nil, repository.ErrNotFound
    }
    return r.data[recordType][recordID], nil
}
```

Finally, let's implement a Put function inside it, as shown:

```
// Put adds a rating for a given record.
func (r *Repository) Put(ctx context.Context, recordID model.
RecordID, recordType model.RecordType, rating *model.Rating)
error {
    if _, ok := r.data[recordType]; !ok {
        r.data[recordType] = map[model.RecordID][]model.
Rating{}
    }
    r.data[recordType][recordID] =
append(r.data[recordType][recordID], *rating)
    return nil
}
```

The preceding implementation is using a nested map to store all records inside it. If we didn't define separate types, RatingID, RatingType, and UserID, it would be harder to understand the types of the keys in the map because we would be using primitives such as string and int, which are less self-descriptive.

Controller

Let's add a controller for our rating service. In the rating/internal/controller/rating package, create a controller.go file:

```
package rating

import (
```

```
    "context"
    "errors"

    "movieexample.com/rating/internal/repository"
    "movieexample.com/rating/pkg/model"
)


// ErrNotFound is returned when no ratings are found for a
// record.
var ErrNotFound = errors.New("ratings not found for a record")

type ratingRepository interface {
    Get(ctx context.Context, recordID model.RecordID,
recordType model.RecordType) ([]model.Rating, error)
    Put(ctx context.Context, recordID model.RecordID,
recordType model.RecordType, rating *model.Rating) error
}


// Controller defines a rating service controller.
type Controller struct {
    repo ratingRepository
}


// New creates a rating service controller.
func New(repo ratingRepository) *Controller {
    return &Controller{repo}
}
```

Let's add functions for writing and getting an aggregated rating:

```
// GetAggregatedRating returns the aggregated rating for a
// record or ErrNotFound if there are no ratings for it.
func (c *Controller) GetAggregatedRating(ctx context.Context,
recordID model.RecordID, recordType model.RecordType) (float64,
error) {
    ratings, err := c.repo.Get(ctx, recordID, recordType)
    if err != nil && err == repository.ErrNotFound {
```

```
            return 0, ErrNotFound
    } else if err != nil {
            return 0, err
    }
    sum := float64(0)
    for _, r := range ratings {
            sum += float64(r.Value)
    }
    return sum / float64(len(ratings)), nil
}


// PutRating writes a rating for a given record.
func (c *Controller) PutRating(ctx context.Context, recordID
model.RecordID, recordType model.RecordType, rating *model.
Rating) error {
    return c.repo.Put(ctx, recordID, recordType, rating)
}
```

In this example, it is easy to see how the controller logic is different from the repository one. The repository provides an interface to get all ratings for a record and the controller implements the aggregation logic for them.

Handler

Let's implement the service handler in the `rating/internal/handler/http/http.go` file, using the following code:

```
package http

import (
    "encoding/json"
    "errors"
    "log"
    "net/http"
    "strconv"

    "movieexample.com/rating/internal/controller"
    "movieexample.com/rating/pkg/model"
)
```

```go
// Handler defines a rating service controller.
type Handler struct {
    ctrl *rating.Controller
}


// New creates a new rating service HTTP handler.
func New(ctrl *rating.Controller) *Handler {
    return &Handler{ctrl}
}
```

Now, let's add a function for handling HTTP requests to our service:

```go
// Handle handles PUT and GET /rating requests.
func (h *Handler) Handle(w http.ResponseWriter, req *http.
Request) {
    recordID := model.RecordID(req.FormValue("id"))
    if recordID == "" {
        w.WriteHeader(http.StatusBadRequest)
        return
    }
    recordType := model.RecordType(req.FormValue("type"))
    if recordType == "" {
        w.WriteHeader(http.StatusBadRequest)
        return
    }
    switch req.Method {
    case http.MethodGet:
        v, err := h.ctrl.GetAggregatedRating(req.Context(),
recordID, recordType)
        if err != nil && errors.Is(err, rating.ErrNotFound) {
            w.WriteHeader(http.StatusNotFound)
            return
        }
        if err := json.NewEncoder(w).Encode(v); err != nil {
            log.Printf("Response encode error: %v\n", err)
        }
```

```
    case http.MethodPut:
        userID := model.UserID(req.FormValue("userId"))
        v, err := strconv.ParseFloat(req.FormValue("value"),
64)
        if err != nil {
            w.WriteHeader(http.StatusBadRequest)
            return
        }
        if err := h.ctrl.PutRating(req.Context(), recordID,
recordType, &model.Rating{UserID: userID, Value: model.
RatingValue(v)}); err != nil {
            log.Printf("Repository put error: %v\n", err)
            w.WriteHeader(http.StatusInternalServerError)
        }
    default:
        w.WriteHeader(http.StatusBadRequest)
    }
}
```

The handler we implemented handles both GET and PUT requests. Note the way we handle some special cases such as an empty id value from the request—in that case, we return a special error code, http.StatusBadRequest, indicating that the API request was invalid. If the record is not found, we return http.StatusNotFound, and if we encounter any unexpected errors when accessing our database, we return http.StatusInternalServerError.

Using such standard HTTP error codes helps the clients to differentiate between the types of errors and implement the logic for detecting and correctly handling such issues.

Main

Let's write the main file for our service. In rating/cmd/main.go, write the following logic:

```
package main

import (
    "log"
    "net/http"

    "movieexample.com/rating/internal/controller/rating"
    httphandler "movieexample.com/rating/internal/handler/http"
```

```
        "movieexample.com/rating/internal/repository/memory"
)

func main() {
    log.Println("Starting the rating service")
    repo := memory.New()
    ctrl := rating.New(repo)
    h := httphandler.New(ctrl)
    http.Handle("/rating", http.HandlerFunc(h.Handle))
    if err := http.ListenAndServe(":8082", nil); err != nil {
        panic(err)
    }
}
```

The `main` function we created is similar to the main function of the metadata service; it initializes all components of a service and starts an HTTP handler.

Now, we are ready to implement our last service.

Movie service

Let's summarize the logic of the movie service:

- **API**: Get the details for a movie, including the aggregated movie rating and movie metadata.
- **Database**: None.
- **Interacts with services**: Movie metadata and rating.
- **Data model type**: Movie details.

This logic would translate into the following packages:

- `cmd`: Contains the main function for starting the service
- `controller`: Our service logic (read rating and metadata)
- `gateway`: Logic for calling the other services
- `handler`: API handler for a service

The directory structure is as follows:

- `movie/cmd`
- `movie/internal/controller`

- `movie/internal/gateway`
- `movie/internal/handler`
- `movie/pkg`

Once you have created these directories, let's move on to the implementation of the service.

Model

Create a `model.go` file in the `movie/pkg/model` directory and write the following logic:

```
package model

import "movieexample.com/metadata/pkg/model"

// MovieDetails includes movie metadata its aggregated
// rating.
type MovieDetails struct {
    Rating   *float64    `json:"rating,omitEmpty"`
    Metadata model.Metadata `json:"metadata`
}
```

Note that the file imports the model package of a metadata service containing the `Metadata` structure that we can reuse in our service.

Gateways

In the previous examples, the services did not interact with each other and just provided an API for this. The movie service won't access any database by itself but instead is going to interact with both the movie metadata and the rating service.

Let's create the logic for interacting with both services.

First, let's create an error that we are going to use in our gateways. In the `movie/internal/gateway` package, create an `error.go` file, using the following code block:

```
package gateway

import "errors"

// ErrNotFound is returned when the data is not found.
var ErrNotFound = errors.New("not found")
```

Now, let's write an HTTP gateway for the movie metadata service. In the `movie/gateway/metadata/http` directory, create a `metadata.go` file:

```go
package http

import (
    "context"
    "encoding/json"
    "fmt"
    "net/http"

    "movieexample.com/metadata/pkg/model"
    "movieexample.com/movie/internal/gateway"
)

// Gateway defines a movie metadata HTTP gateway.
type Gateway struct {
    addr string
}

// New creates a new HTTP gateway for a movie metadata
// service.
func New(addr string) *Gateway {
    return &Gateway{addr}
}
```

Let's implement a `Get` function in it:

```go
// Get gets movie metadata by a movie id.
func (g *Gateway) Get(ctx context.Context, id string) (*model.Metadata, error) {
    req, err := http.NewRequest(http.MethodGet, g.addr+"/metadata", nil)
    if err != nil {
        return nil, err
    }
    req = req.WithContext(ctx)
    values := req.URL.Query()
```

```
    values.Add("id", id)
    req.URL.RawQuery = values.Encode()
    resp, err := http.DefaultClient.Do(req)
    if err != nil {
        return nil, err
    }
    defer resp.Body.Close()
    if resp.StatusCode == http.StatusNotFound {
        return nil, gateway.ErrNotFound
    } else if resp.StatusCode/100 != 2 {
        return nil, fmt.Errorf("non-2xx response: %v", resp)
    }
    var v *model.Metadata
    if err := json.NewDecoder(resp.Body).Decode(&v); err != nil
{

        return nil, err
    }
    return v, nil
}
```

Now, let's write an HTTP gateway for the rating service. In the `movie/gateway/rating/http` directory, create a `rating.go` file:

```
package http

import (
    "context"
    "encoding/json"
    "fmt"
    "net/http"

    "movieexample.com/movie/internal/gateway"
    "movieexample.com/rating/pkg/model"
)

// Gateway defines an HTTP gateway for a rating service.
type Gateway struct {
```

```
    addr string
}

// New creates a new HTTP gateway for a rating service.
func New(addr string) *Gateway {
    return &Gateway{addr}
}
```

Let's add logic for getting the aggregated rating:

```
// GetAggregatedRating returns the aggregated rating for a
// record or ErrNotFound if there are no ratings for it.
func (g *Gateway) GetAggregatedRating(ctx context.Context,
recordID model.RecordID, recordType model.RecordType) (float64,
error) {
    req, err := http.NewRequest(http.MethodGet, g.addr+"/
rating", nil)
    if err != nil {
        return 0, err
    }
    req = req.WithContext(ctx)
    values := req.URL.Query()
    values.Add("id", string(recordID))
    values.Add("type", fmt.Sprintf("%v", recordType))
    req.URL.RawQuery = values.Encode()
    resp, err := http.DefaultClient.Do(req)
    if err != nil {
        return 0, err
    }
    defer resp.Body.Close()
    if resp.StatusCode == http.StatusNotFound {
        return 0, gateway.ErrNotFound
    } else if resp.StatusCode/100 != 2 {
        return 0, fmt.Errorf("non-2xx response: %v", resp)
    }
    var v float64
    if err := json.NewDecoder(resp.Body).Decode(&v); err != nil
{
```

```
            return 0, err
        }
        return v, nil
}
```

Finally, let's add a function for handling a rating creation request:

```
// PutRating writes a rating.
func (g *Gateway) PutRating(ctx context.Context, recordID
model.RecordID, recordType model.RecordType, rating *model.
Rating) error {
        req, err := http.NewRequest(http.MethodPut, g.addr+"/
rating", nil)
        if err != nil {
            return err
        }
        req = req.WithContext(ctx)
        values := req.URL.Query()
        values.Add("id", string(recordID))
        values.Add("type", fmt.Sprintf("%v", recordType))
        values.Add("userId", string(rating.UserID))
        values.Add("value", fmt.Sprintf("%v", rating.Value))
        req.URL.RawQuery = values.Encode()
        resp, err := http.DefaultClient.Do(req)
        if err != nil {
            return err
        }
        defer resp.Body.Close()
        if resp.StatusCode/100 != 2 {
            return fmt.Errorf("non-2xx response: %v", resp)
        }
        return nil
}
```

At this point, we have both gateways and can implement the controller aggregating the data from them.

Controller

In the `movie/internal/controller/movie` directory, create a `controller.go` file:

```
package movie

import (
    "context"
    "errors"

    metadatamodel "movieexample.com/metadata/pkg/model"
    "movieexample.com/movie/internal/gateway"
    "movieexample.com/movie/pkg/model"
    ratingmodel "movieexample.com/rating/pkg/model"
)

// ErrNotFound is returned when the movie metadata is not
// found.
var ErrNotFound = errors.New("movie metadata not found")
```

Let's define the interfaces for the services we will be calling:

```
type ratingGateway interface {
    GetAggregatedRating(ctx context.Context, recordID
ratingmodel.RecordID, recordType ratingmodel.RecordType)
(float64, error)
    PutRating(ctx context.Context, recordID ratingmodel.
RecordID, recordType ratingmodel.RecordType, rating
*ratingmodel.Rating) error
}

type metadataGateway interface {
    Get(ctx context.Context, id string) (*metadatamodel.
Metadata, error)
}
```

Now, we can define our service controller:

```
// Controller defines a movie service controller.
type Controller struct {
```

```
    ratingGateway    ratingGateway
    metadataGateway metadataGateway
}


// New creates a new movie service controller.
func New(ratingGateway ratingGateway, metadataGateway
metadataGateway) *Controller {
    return &Controller{ratingGateway, metadataGateway}
}
```

Finally, let's implement the function for getting the movie details, including both its rating and metadata:

```
// Get returns the movie details including the aggregated
// rating and movie metadata.
// Get returns the movie details including the aggregated
rating and movie metadata.
func (c *Controller) Get(ctx context.Context, id string)
(*model.MovieDetails, error) {
    metadata, err := c.metadataGateway.Get(ctx, id)
    if err != nil && errors.Is(err, gateway.ErrNotFound) {
        return nil, ErrNotFound
    } else if err != nil {
        return nil, err
    }
    details := &model.MovieDetails{Metadata: *metadata}
    rating, err := c.ratingGateway.GetAggregatedRating(ctx,
ratingmodel.RecordID(id), ratingmodel.RecordTypeMovie)
    if err != nil && !errors.Is(err, gateway.ErrNotFound) {
        // Just proceed in this case, it's ok not to have
ratings yet.
    } else if err != nil {
        return nil, err
    } else {
        details.Rating = &rating
    }
    return details, nil
}
```

Note that we redefine `ErrNotFound` in different components. While we could have just exported it to some shared package, sometimes it is better to keep it independent. Otherwise, we may confuse one error for another (for example, rating not found or metadata not found).

Handler

In the `movie/internal/handler/http` package, add the `http.go` file, using the following logic:

```go
package http

import (
    "encoding/json"
    "errors"
    "log"
    "net/http"

    "movieexample.com/movie/internal/controller/movie"
)

// Handler defines a movie handler.
type Handler struct {
    ctrl *movie.Controller
}

// New creates a new movie HTTP handler.
func New(ctrl *movie.Controller) *Handler {
    return &Handler{ctrl}
}

// GetMovieDetails handles GET /movie requests.
func (h *Handler) GetMovieDetails(w http.ResponseWriter, req
*http.Request) {
    id := req.FormValue("id")
    details, err := h.ctrl.Get(req.Context(), id)
    if err != nil && errors.Is(err, movie.ErrNotFound) {
        w.WriteHeader(http.StatusNotFound)
        return
    } else if err != nil {
```

```
            log.Printf("Repository get error: %v\n", err)
            w.WriteHeader(http.StatusInternalServerError)
            return
        }
        if err := json.NewEncoder(w).Encode(details); err != nil {
            log.Printf("Response encode error: %v\n", err)
        }
    }
```

Now, we are finally ready to write a main file for the movie service.

Main file

In the movie/cmd package, create a main.go file, using the following code block:

```
package main

import (
    "log"
    "net/http"

    "movieexample.com/movie/internal/controller/movie"
    metadatagateway "movieexample.com/movie/internal/gateway/
metadata/http"
    ratinggateway "movieexample.com/movie/internal/gateway/
rating/http"
    httphandler "movieexample.com/movie/internal/handler/http"
)

func main() {
    log.Println("Starting the movie service")
    metadataGateway := metadatagateway.New("localhost:8081")
    ratingGateway := ratinggateway.New("localhost:8082")
    ctrl := movie.New(ratingGateway, metadataGateway)
    h := httphandler.New(ctrl)
    http.Handle("/movie", http.HandlerFunc(h.GetMovieDetails))
    if err := http.ListenAndServe(":8083", nil); err != nil {
        panic(err)
```

```
        }
    }
```

At this point, we have the logic for all three services. Note that we used static service addresses, `localhost:8081`, `localhost:8082`, and `localhost:8083`, in this example. This allows you to run the services locally; however, this would not work if we deployed our services to the cloud or any other deployment platform. In the next chapter, we are going to cover this aspect and continue improving our microservices. You can run the services we just created by executing this command inside the `cmd` directory of each service:

```
go run *.go
```

Then, you can call the metadata service API using the following command:

```
curl localhost:8081?id=1
```

You can call the rating service API using a similar command:

```
curl localhost:8082?id=1&type=2
```

Finally, you can call the movie service using the following command:

```
curl localhost:8083?id=1
```

All of the preceding requests should return an HTTP 404 error, indicating that records are not found—we do not have any data yet, so this is expected.

At this point, we have illustrated how to bootstrap and manually test our example microservices and are ready to move on to the next chapter.

Summary

In this section, we have covered lots of topics, including the most important recommendations for writing Go applications and the standards for the project layout of Go applications. The knowledge we gained helped us during the code scaffolding of our microservices—we have tried to implement our microservice code in an idiomatic way as much as possible.

You have also learned how to split each of your microservices into multiple layers, each responsible for its own logic. We have illustrated how to separate the business logic from the code accessing the database, and how to separate the API handler logic from both, as well as from the logic performing remote calls between the services.

While the amount of information in this chapter is quite overwhelming, we have made a solid start and are ready to move on to more advanced topics. In the next chapter, we are going to see how the microservices we created can explore each other, so we can finally test them.

Further reading

- *Effective Go*: https://go.dev/doc/effective_go
- *Go Code Review Comments*: https://github.com/golang/go/wiki/CodeReviewComments
- *Project layout*: https://github.com/golang-standards/project-layout
- *Package names*: https://go.dev/blog/package-names

Service Discovery

In the previous chapter, we created our example microservices and let them communicate with each other, using static local addresses hardcoded into each service. This approach would work until we to add or remove service instances dynamically, known as service discovery – letting microservices find each other in a dynamic environment. Setting up service discovery is the first step for writing and preparing scalable microservices in a real production environment.

In this chapter, we are going to cover the following topics:

- Service discovery overview
- Service discovery solutions
- Adopting service discovery

We will use the microservices we created in the previous chapter to illustrate how to use the service discovery solutions. Now, let's move on to the overview of the service discovery concepts.

Technical requirements

In order to complete this chapter, you need Go 1.11 or above. Additionally, you will need a Docker tool, which you can download from `https://www.docker.com`.

You can find the source files for this chapter on GitHub: `https://github.com/PacktPublishing/microservices-with-go/tree/main/Chapter03`.

Service discovery overview

In the previous chapter, we created an application consisting of three microservices. The relationship between the services is illustrated in the following diagram:

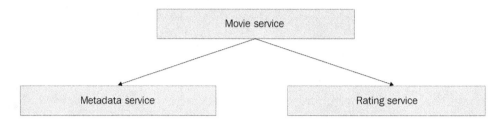

Figure 3.1 – Relationship between our microservices

As you can see, the movie service calls both the metadata and rating service for fetching the complete movie details.

But how would our services send requests? How would they know the addresses of each other?

In our example, we used pre-programmed static values for the API handlers. The settings we used were as follows:

- **Metadata service**: `localhost:8081`

- **Rating service**: `localhost:8082`

- **Movie service**: `localhost:8083`

In our approach, each service would need to know the exact address of the other services it would communicate with. This approach would work until we more than one instance of each microservice. In this case, we would have multiple challenges:

- What address should you use when you have multiple instances?

- How to handle the situation when some instances become unavailable?

The first problem is relatively easy to solve if you have a static set of instances – your services would need to keep a list of addresses for each service they need to call. However, this approach is not flexible for the following reasons:

- Each time you need to add or remove instances, you need to update the configuration of each calling service.

- If an instance becomes unavailable for an extended period (for example, due to network failure), your services would still keep calling it until you updated their configuration.

How do you solve these problems properly?

The problem we just described for our microservices is called **service discovery**. In general, service discovery addresses multiple problems, as follows:

- How to discover the instance(s) of a particular service

- How to add and remove instances of service in a discoverable environment

- How to handle the issue when instances become unresponsive

Let's see how each of these features would work.

Registry

The foundation of service discovery is a registry (also known as a service registry), which stores information about available service instances. It has the following features:

- Register an instance of a service.

- Deregister an instance of a service.

- Return the list of all the instances of the service in the form of their network addresses.

This is an example of service discovery registry data:

Service name	Address list
Movie service	172.18.10.2:2520 172.18.12.55:8800 172.18.89.10:2450
Rating service	172.18.25.11:1100 172.18.9.55:2830
Movie metadata service	172.18.79.115:3512 172.17.3.8:9900

Table 3.1 – Registry data

Each service can either register in the registry by itself or use some library or tool to register on service startup automatically. Once the service is registered, it starts being monitored via health checks to ensure the registry contains only available instances.

Let's now see the two common models of adopting service discovery.

Service discovery models

For the applications, there are two ways of interacting with the registry:

- **Client-side service discovery**: Access the registry directly from the application using a registry client.

- **Server-side service discovery**: Access the registry indirectly via a load balancer, a special server that forwards requests to available instances.

Let's see the advantages and disadvantages of each model.

Client-side service discovery

In the client-side service discovery model, each application or service accesses the service registry directly by requesting all available instances of a target service. When the application receives a response, it uses the addresses of the target service for making requests. The logic is illustrated in the following diagram:

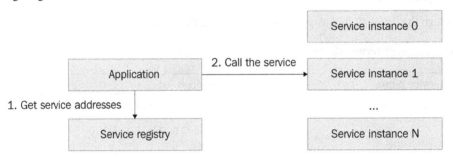

Figure 3.2 – Client-side service discovery

In this model, the application is responsible for balancing the load on the service it is calling – if an application picked just one instance from the list and kept calling it all the time, it would overload that instance and underutilize the other ones.

The downside of this model is that the calling application needs to be programmed with load-balancing logic. In addition, this couples service discovery and load-balancing logic with the application code, making the application more complex.

Server-side service discovery

The server-side service discovery model adds an extra layer to the interaction between the calling applications and the registry. Instead of calling the registry directly, applications send their requests to target microservices via a special server called a load balancer. The load balancer is responsible for interacting with the registry and distributing requests between all available instances.

The following diagram will help you understand the server-side service discovery model:

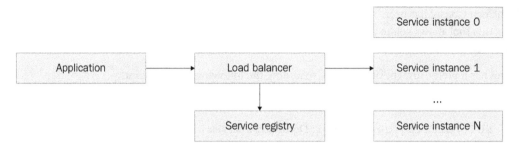

Figure 3.3 – Server-side service discovery

In the diagram, the application calls the target service via a load balancer, accessing the service registry to read the list of active service instances from it. In this model, the application does not need to know about the registry. This is the main benefit of the server-side service discovery model: it helps decouple the interaction with the service registry from each calling application, making the application logic simpler. The downside of the model is that it requires setting up and using a load balancer. The latter is rather an untrivial operation, and we are not going to cover it in this book.

Now, let's see how the registry can keep the list of only active instances of each service.

Service health monitoring

The registry keeps the information about the instances up to date either via a pull or a push model:

- **Pull model**: The service registry periodically performs a health check for each known instance.
- **Push model**: The application renews its status by contacting the registry.

The pull model removes the need to implement status renewal on a service level. In a push model, the application is responsible for renewing its status or telling the service registry about its healthy status.

Now, as we have covered the theoretical basics of service discovery, let's see which existing solutions you can use to enable your microservices.

Service discovery solutions

In this section, we are going to describe the existing service discovery solutions available for your use – **HashiCorp Consul** and **Kubernetes**. Then, you will learn about the most popular tools that can be used by microservice developers to perform service discovery.

HashiCorp Consul

HashiCorp Consul has been a pretty popular solution for service discovery for many years. Written in Go, this tool allows you to set up service discovery for your services and applications quite easily, using its clients or API.

Consul has a pretty straightforward API, including the following key endpoints:

- `PUT /catalog/register`: Register a service instance.
- `PUT /catalog/deregister`: Deregister a service instance.
- `GET /catalog/services`: Get the available instances of a service.

Client applications can access the Consul catalog either via the API or in server-side service discovery mode, using a DNS service.

You can learn more about Consul by checking the official website: `https://consul.io`.

Kubernetes

Kubernetes is a popular open source platform for running, scaling, and managing collections of applications, such as microservices.

One of the features of Kubernetes is the ability to register and discover the services that are running in it. Kubernetes provides an API for retrieving the list of network addresses of each service that is being updated automatically, so users can use it in client-side discovery mode. Alternatively, it allows its users to plug in a load balancer to use it for server-side discovery instead.

We are going to cover Kubernetes later, in *Chapter 8* of this book. Now, let's see how we can add service discovery to the applications we created in the previous chapter.

Adopting service discovery

In this section, we are going to illustrate how you can start using service discovery for your applications. We will be using the microservices we created in the previous chapter as an example. Then, you will learn how to add the logic responsible for service discovery to your microservice code.

When you consider enabling service discovery for your services, you have multiple questions to answer, such as the following:

- Which model would you prefer to use – client-side or server-side discovery?
- Which platform will you use for the deployment and orchestration of your microservices?

Answering the second question may already give you a solution – various deployment platforms, including Kubernetes, as well as popular clouds such as AWS, offer service discovery for your services.

If you don't know which deployment platform you are going to use for your services and you are new to microservice development, you may consider client-side service discovery. The client-side discovery model is slightly simpler because your service directly coordinates with the service registry. Later, you will be able to switch to server-side service discovery, if you want.

Let's start preparing your applications for adding service discovery logic.

Preparing the application

Let's list what we want to achieve from our service discovery code.

- The ability to register a service that we are going to use on service startup
- The ability to deregister a service that we are going to use on service shutdown
- The ability to get a list of addresses of a particular service that we are going to use for making the calls to the other services.
- Set up service health monitoring so the service registry is able to remove inactive service instances

It would be great if our service discovery logic weren't directly tied to a particular tool. It is usually a good practice to abstract the actual technology with a more generic interface, which allows us to swap implementations. We can illustrate this using an example – imagine we are using the Hashicorp Consul library, which returns a list of service addresses in the following form:

```
func Service(string, string) ([]*consul.ServiceEntry, *consul.
QueryMeta, error)
```

If we expose these Consul structures in our code and pass these structures around our code base, our service code will be heavily tied to Consul. If we ever decide to switch to another service discovery tool, we would need to replace not only the service discovery implementation logic but also all the code in which it is used.

Instead, let's define a more generic and technology-agnostic interface. For providing the list of service instances, we can just return the list of URLs in the `[] string` format.

The complete interface for our service discovery logic would be as follows:

```
// registry defines a service registry.
type Registry interface {
    // Register creates a service instance record in the
registry.
    Register(ctx context.Context, instanceID string,
serviceName string, hostPort string) error
    // Deregister removes a service instance record from the
registry.
```

```
    Deregister(ctx context.Context, instanceID string,
serviceName string) error
    // ServiceAddresses returns the list of addresses of active
instances of the given service.
    ServiceAddresses(ctx context.Context, serviceID string) ([]
string, error)
    // ReportHealthyState is a push mechanism for reporting
healthy state to the registry.
    ReportHealthyState(instanceID string, serviceName string)
error
}
```

As you may notice, the interface is pretty generic, yet it allows you to create multiple implementations based on different technologies if needed.

You may also notice that the interface includes a `ReportHealthyState` function for reporting the healthy state of a service instance. This function allows us to implement the push-based service health monitoring that we mentioned before, so each microservice would periodically report its health to the service registry. The registry would then be able to remove inactive instances of each service if they don't report a healthy state within some defined interval of time (we are going to assume the interval is 5 seconds in our implementation).

Let's now think about where we would store the service discovery logic for our microservices. I suggest using a package that all three services can access – let's create it in the root directory of our application, under the `pkg` folder. We can call it `pkg/discovery`. Inside it, add a `discovery.go` file and add the following code to it:

```
package discovery

import (
    "context"
    "errors"
    "fmt"
    "math/rand"
    "time"
)

// Registry defines a service registry.
type Registry interface {
    // Register creates a service instance record in the
    // registry.
```

```
    Register(ctx context.Context, instanceID string,
serviceName string, hostPort string) error
    // Deregister removes a service insttance record from
    // the registry.
    Deregister(ctx context.Context, instanceID string,
serviceName string) error
    // ServiceAddresses returns the list of addresses of
    // active instances of the given service.
    ServiceAddresses(ctx context.Context, serviceID string) ([]
string, error)
    // ReportHealthyState is a push mechanism for reporting
    // healthy state to the registry.
    ReportHealthyState(instanceID string, serviceName string)
error
}


// ErrNotFound is returned when no service addresses are
// found.
var ErrNotFound = errors.New("no service addresses found")


// GenerateInstanceID generates a pseudo-random service
// instance identifier, using a service name
// suffixed by dash and a random number.
func GenerateInstanceID(serviceName string) string {
    return fmt.Sprintf("%s-%d", serviceName, rand.New(rand.
NewSource(time.Now().UnixNano())).Int())
}
```

In the code that we just added, we have defined a Registry interface for the service registry. Additionally, we defined the ErrNotFound error that would be returned from the ServiceAddresses function if no active service addresses were found. Finally, we created a GenerateInstanceID function that would help us to generate randomized instance identifiers for use with the Register and Deregister functions.

We are ready to begin our work on its implementation.

Implementing the discovery logic

One of the benefits of the interface that we defined earlier is that we can create multiple implementations and use them in our application. So, for example, we can create one implementation that we can later use in tests, whereas another implementation would be used in production. To illustrate this approach, we are going to create two implementations:

- **In-memory service discovery**: Use an in-memory registry for storing the set of addresses.
- **Consul-based service discovery**: Use the Hashicorp Consul service registry for storing and retrieving service addresses.

Now, let's proceed to implement the logic.

In-memory implementation

Let's start with the in-memory implementation first. In this implementation, we will be storing the service registry records in memory using a simple map data structure. Here are the steps:

1. Create a `pkg/discovery/memory`package file and a `memory.go` file, then add the following:

    ```
    package memory

    import (
        "context"
        "errors"
        "net"
        "sync"
        "time"

        "movieexample.com/pkg/discovery"
    )

    type serviceName string
    type instanceID string

    // Registry defines an in-memory service registry.
    type Registry struct {
        sync.RWMutex
        serviceAddrs map[serviceName]
    map[instanceID]*serviceInstance
    ```

```
    }

    type serviceInstance struct {
        hostPort   string
        lastActive time.Time
    }

    // NewRegistry creates a new in-memory service
    // registry instance.
    func NewRegistry() *Registry {
        return &Registry{serviceAddrs: map[serviceName]
    map[instanceID]*serviceInstance{}}
    }
```

2. Let's implement our `Register` and `Deregister` functions:

```
    // Register creates a service record in the registry.
    func (r *Registry) Register(ctx context.Context,
    instanceID string, serviceName string, hostPort string)
    error {
        r.Lock()
        defer r.Unlock()
        if _, ok := r.serviceAddrs[serviceName]; !ok {
            r.serviceAddrs[serviceName] =
    map[string]*serviceInstance{}
        }
        r.serviceAddrs[serviceName][instanceID] =
    &serviceInstance{hostPort: hostPort, lastActive: time.
    Now()}
        return nil
    }

    // Deregister removes a service record from the
    // registry.
    func (r *Registry) Deregister(ctx context.Context,
    instanceID string, serviceName string) error {
        r.Lock()
        defer r.Unlock()
        if _, ok := r.serviceAddrs[serviceName]; !ok {
```

```
        return nil
    }
    delete(r.serviceAddrs[serviceName], instanceID)
    return nil
}
```

3. Finally, let's implement the remaining two functions of the Registry interface:

```
// ReportHealthyState is a push mechanism for
// reporting healthy state to the registry.
func (r *Registry) ReportHealthyState(instanceID string,
serviceName string) error {
    r.Lock()
    defer r.Unlock()
    if _, ok := r.serviceAddrs[serviceName]; !ok {
        return errors.New("service is not registered
yet")
    }
    if _, ok := r.serviceAddrs[serviceName][instanceID];
!ok {
        return errors.New("service instance is not
registered yet")
    }
    r.serviceAddrs[serviceName][instanceID].lastActive =
time.Now()
    return nil
}


// ServiceAddresses returns the list of addresses of
// active instances of the given service.
func (r *Registry) ServiceAddresses(ctx context.Context,
serviceName string) ([]string, error) {
    r.RLock()
    defer r.RUnlock()
    if len(r.serviceAddrs[serviceName]) == 0 {
        return nil, discovery.ErrNotFound
    }
    var res []string
```

```go
    for _, i := range r.serviceAddrs[serviceName] {
        if i.lastActive.Before(time.Now().Add(-5 * time.
Second)) {
            continue
        }
        res = append(res, i.hostPort)
    }
    return res, nil
}
```

This implementation can be used in tests or simple applications running on a single server. The implementation is based on a combination of a map data structure and `sync.RWMutex`, allowing reads and writes to the map concurrently. In the map, we store `serviceInstance` structures containing the instance address and the last time of a successful health check for it, which can be set by calling a `ReportHealthyState` function. In the `ServiceAddresses` function, we only return instances with successful health checks from within the last 5 seconds.

Let's now move to the Consul-based service registry implementation.

Consul-based implementation

The implementation that we are going to work on now will use Hashicorp Consul as a service registry:

1. First, create a `pkg/discovery/consul` package and add to it a file named `consul.go`:

    ```go
    package consul

    import (
        "context"
        "errors"
        "fmt"
        "strconv"
        "strings"

        consul "github.com/hashicorp/consul/api"
        "movieexample.com/pkg/discovery"
    )

    // Registry defines a Consul-based service regisry.
    type Registry struct {
    ```

```
        client *consul.Client
}


// NewRegistry creates a new Consul-based service
// registry instance.
func NewRegistry(addr string) (*Registry, error) {
    config := consul.DefaultConfig()
    config.Address = addr
    client, err := consul.NewClient(config)
    if err != nil {
        return nil, err
    }
    return &Registry{client: client}, nil
}
```

2. Now, let's implement the functions of our interface to register and reregister the records:

```
 // Register creates a service record in the registry.
func (r *Registry) Register(ctx context.Context,
instanceID string, serviceName string, hostPort string)
error {
    parts := strings.Split(hostPort, ":")
    if len(parts) != 2 {
        return errors.New("hostPort must be in a form of
<host>:<port>, example: localhost:8081")
    }
    port, err := strconv.Atoi(parts[1])
    if err != nil {
        return err
    }
    return r.client.Agent().ServiceRegister(&consul.
AgentServiceRegistration{
        Address: parts[0],
        ID:        instanceID,
        Name:      serviceName,
        Port:      port,
        Check:     &consul.AgentServiceCheck{CheckID:
instanceID, TTL: "5s"},
```

```
        })
    }


    // Deregister removes a service record from the
    // registry.
    func (r *Registry) Deregister(ctx context.Context,
    instanceID string, _ string) error {
        return r.client.Agent().ServiceDeregister(instanceID)
    }
```

3. Finally, let's implement the remaining registry functions:

```
    // ServiceAddresses returns the list of addresses of
    // active instances of the given service.
    func (r *Registry) ServiceAddresses(ctx context.Context,
    serviceName string) ([]string, error) {
        entries, _, err := r.client.Health().
    Service(serviceName, "", true, nil)
        if err != nil {
            return nil, err
        } else if len(entries) == 0 {
            return nil, discovery.ErrNotFound
        }
        var res []string
        for _, e := range entries {
            res = append(res, res = append(res, fmt.
    Sprintf("%s:%d", e.Service.Address, e.Service.Port)))
        }
        return res, nil
    }


    // ReportHealthyState is a push mechanism for
    // reporting healthy state to the registry.
    func (r *Registry) ReportHealthyState(instanceID string,
    _ string) error {
        return r.client.Agent().PassTTL(instanceID, "")
    }
```

Our client depends on an external library, github.com/hashicorp/consul/api. We need to fetch it now by running go mod tidy inside our src directory. After this, Go should fetch the dependency, and our logic should be able to compile.

Now, we are ready to plug the logic we just created into our microservices.

Using the discovery logic

Now, we need to add logic for initializing and discovering the services. Currently, only the movie service communicates with the other two, so we are going to illustrate how to add service discovery using the movie service as an example.

Let's start with our gateways:

1. In the previous chapter, we created two gateways for calling the metadata and the rating services. Let's modify their structures to the one shown here:

    ```
    type Gateway struct {
        registry discovery.Registry
    }
    ```

2. Also, change the New function format to the following:

    ```
    func New(registry discovery.Registry) *Gateway {
        return &Gateway{registry}
    }
    ```

3. Now, the gateways require a registry on creation. We can change the beginning of the Get function of the metadata gateway to this now:

    ```
    func (g *Gateway) Get(ctx context.Context, id string)
    (*model.Metadata, error) {
        addrs, err := g.registry.ServiceAddresses(ctx,
    "metadata")
        if err != nil {
            return nil, err
        }
        url := "http://" + addrs[rand.Intn(len(addrs))] + "/
    metadata"
        log.Printf("Calling metadata service. Request: GET "
    + url)
        req, err := http.NewRequest(http.MethodGet, url, nil)
    ```

You may notice that instead of calling a static pre-configured address, we now first get the available addresses of the metadata from the registry. This is the essence of service discovery – we use the data from the registry for making remote calls between our services. After we get the list of service addresses, we pick a random one using a `rand.Intn` function. By doing this, we balance the load between the active instances, randomly selecting any available instance on each request.

4. Now, updating the rating gateway in the same way, we changed the metadata service.

5. The next step is to update the `main` functions of our services so that each service will register and deregister itself in the service registry. Let's update the metadata service first. Update its `main` function to the following:

```go
const serviceName = "metadata"

func main() {
    var port int
    flag.IntVar(&port, "port", 8081, "API handler port")
    flag.Parse()
    log.Printf("Starting the metadata service on port
%d", port)
    registry, err := consul.NewRegistry("localhost:8500")
    if err != nil {
        panic(err)
    }
    ctx := context.Background()
    instanceID := discovery.
GenerateInstanceID(serviceName)
    if err := registry.Register(ctx, instanceID,
serviceName, fmt.Sprintf("localhost:%d", port)); err !=
nil {
        panic(err)
    }
    go func() {
        for {
            if err := registry.
ReportHealthyState(instanceID, serviceName); err != nil {
                log.Println("Failed to report healthy
state: " + err.Error())
            }
            time.Sleep(1 * time.Second)
```

```
        }
    }()
    defer registry.Deregister(ctx, instanceID,
serviceName)
    repo := memory.New()
    svc := metadata.New(repo)
    h := httphandler.New(svc)
    http.Handle("/metadata", http.HandlerFunc(h.
GetMetadataByID))
    if err := http.ListenAndServe(fmt.Sprintf(":%d",
port), nil); err != nil {
        panic(err)
    }
}
```

In the preceding code, we added the logic for registering and deregistering the service in the Consul-based service registry and reporting its healthy state to it every second.

6. Let's add similar logic to the rating service. Update its main function as follows:

```
func main() {
    var port int
    flag.IntVar(&port, "port", 8082, "API handler port")
    flag.Parse()
    log.Printf("Starting the rating service on port %d",
port)
    registry, err := consul.NewRegistry("localhost:8500")
    if err != nil {
        panic(err)
    }
    ctx := context.Background()
    instanceID := discovery.
GenerateInstanceID(serviceName)
    if err := registry.Register(ctx, instanceID,
serviceName, fmt.Sprintf("localhost:%d", port)); err !=
nil {
        panic(err)
    }
    go func() {
        for {
```

```
            if err := registry.
ReportHealthyState(instanceID, serviceName); err != nil {
                log.Println("Failed to report healthy
state: " + err.Error())
            }
            time.Sleep(1 * time.Second)
        }
    }()
    defer registry.Deregister(ctx, instanceID,
serviceName)
    repo := memory.New()
    svc := controller.New(repo)
    h := httphandler.New(svc)
    http.Handle("/rating", http.HandlerFunc(h.Handle))
    if err := http.ListenAndServe(fmt.Sprintf(":%d",
port), nil); err != nil {
        panic(err)
    }
}
```

The changes that we just made are similar to the ones we did for the metadata service.

7. The last step is to modify the `main` function of the movie service, replacing it with the following:

```
func main() {
    var port int
    flag.IntVar(&port, "port", 8083, "API handler port")
    flag.Parse()
    log.Printf("Starting the movie service on port %d",
port)
    registry, err := consul.NewRegistry("localhost:8500")
    if err != nil {
        panic(err)
    }
    ctx := context.Background()
    instanceID := discovery.
GenerateInstanceID(serviceName)
    if err := registry.Register(ctx, instanceID,
serviceName, fmt.Sprintf("localhost:%d", port)); err !=
nil {
```

```
        panic(err)
    }
    go func() {
        for {
            if err := registry.
ReportHealthyState(instanceID, serviceName); err != nil {
                log.Println("Failed to report healthy
state: " + err.Error())
            }
            time.Sleep(1 * time.Second)
        }
    }()
    defer registry.Deregister(ctx, instanceID,
serviceName)
    metadataGateway := metadatagateway.New(registry)
    ratingGateway := ratinggateway.New(registry)
    svc := movie.New(ratingGateway, metadataGateway)
    h := httphandler.New(svc)
    http.Handle("/movie", http.HandlerFunc(h.
GetMovieDetails))
    if err := http.ListenAndServe(fmt.Sprintf(":%d",
port), nil); err != nil {
        panic(err)
    }
}
```

At this point, we have successfully added Consul-based service discovery to our applications. Let's illustrate how it works in practice:

1. In order to run our applications now, you would need Hashicorp Consul to run locally. The easiest way would be to run it using a Docker tool. Assuming you have already installed Docker from its website, docker.com, you can run the following command:

```
docker run -d -p 8500:8500 -p 8600:8600/udp --name=dev-consul
consul agent -server -ui -node=server-1 -bootstrap-expect=1
-client=0.0.0.0
```

The preceding command runs Hashicorp Consul inside Docker in development mode, exposing its ports 8500 and 8600 for local use.

2. Run each microservice by executing this command inside each cmd directory:

```
go run *.go
```

3. Now, go to the Consul web UI via its link, http://localhost:8500/. When you open the **Services** tab, you should see the list of our services and an active Consul instance:

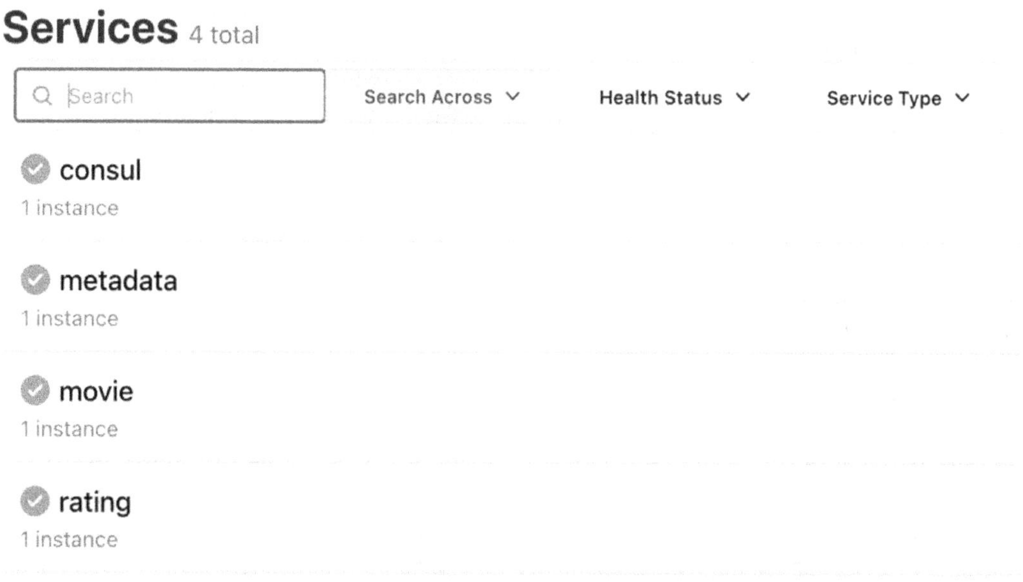

Services 4 total

Q Search		Search Across ∨	Health Status ∨	Service Type ∨

✓ consul
1 instance

✓ metadata
1 instance

✓ movie
1 instance

✓ rating
1 instance

Figure 3.4 – Consul web view of active service instances

You can optionally add some additional instances of each service by running the following:

```
go run *.go --port <PORT>
```

If you run the preceding command, replace the <PORT> placeholder with unique port numbers that are not in use yet (in our examples, we used ports 8081, 8082, and 8083, so you can run with port numbers starting with 8084). The result of each command would be additional healthy instances in the Consul service view illustrated earlier.

You can also try shutting down any service manually by terminating the go run commands and seeing how instances change their states from Passing to Critical.

4. To test API requests, ensure you have at least one healthy instance of each service and make the following request to a movie service:

```
curl -v localhost:8083/movie?id=1
```

5. Check the output logs of the movie service now (you should be able to see them in the terminal where you ran the `go  run` command for the movie service). If you did everything correctly, you should see a similar line:

```
2022/06/08 13:37:42 Calling metadata service. Request:
GET http://localhost:8081/metadata
```

The preceding line is the result of a call to the service registry backed by the Consul. In our metadata service gateway implementation, we select a random active instance from the registry and log its address before making a call. If you have more than one instance of a metadata service, you can make multiple `curl` requests listed previously and see that the movie service always picks a random instance among them.

At this point, we have illustrated how to use service discovery with our microservices. We can now dynamically scale our microservices by adding and removing their instances without any need to change the service code. We also have two working implementations of a service registry that you can use in your code. Now, we are ready to move to the next chapter, covering another important topic, data serialization.

Summary

In this chapter, we did an overview of service discovery and compared its different models. You have learned what the service registry is and what its main service discovery models are. We have illustrated how to use a client-side service discovery model by providing two implementations, one using an in-memory set of data and another using Hashicorp Consul. We have also plugged the Consul-based implementation into our microservices to demonstrate how to use it in the microservice logic. Now, you know how to add and use service discovery in your applications.

In the next chapter, we are going to discuss another important topic: serialization. You will learn how to encode and decode the data transferred between the services. This will help us move to further topics, covering the communication between the services we will cover in *Chapter 5*.

Further reading

- *Service discovery overview*: https://www.nginx.com/blog/service-discovery-in-a-microservices-architecture
- *Server-side service discovery*: https://microservices.io/patterns/server-side-discovery.html

4

Serialization

In previous chapters, we have learned how to scaffold Go microservices, create HTTP API endpoints, and set up service discovery to let our microservices commsunicate with each other. This knowledge already provides a solid foundation for building microservices; however, we are going to continue our journey with more advanced topics.

In this chapter, we will explore **serialization**, a process that allows the encoding and decoding of data for storing or sending between services.

To illustrate how to use serialization, we are going to define data structures transferred between the services using the **Protocol Buffers** format, which is widely used across the industry and has a simple syntax, as well as very size-efficient encoding.

Finally, we are going to illustrate how you can generate code for the Protocol Buffers structures and demonstrate how efficient Protocol Buffers encoding is compared to some other formats, such as XML and JSON.

In this chapter, we are going to cover the following topics:

- The basics of serialization
- Using Protocol Buffers
- Best practices with serialization

Now, let's continue to the basics of serialization.

Technical requirements

To complete this chapter, you will need to have Go 1.11 or above and the Protocol Buffers compiler. We will be using the official Protocol Buffers compiler; you can install it by running the following:

```
go install google.golang.org/protobuf/cmd/protoc-gen-go@latest
export PATH="$PATH:$(go env GOPATH)/bin"
```

You can find the code examples for this chapter on GitHub at the following link:

```
https://github.com/PacktPublishing/microservices-with-go/tree/main/
Chapter04
```

The basics of serialization

Serialization is the process of converting data into a format that allows you to transfer it, store it, and later deconstruct it back.

This process is illustrated in the following diagram:

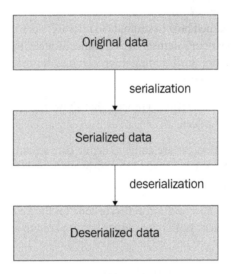

Figure 4.1 – The serialization and deserialization process

As illustrated in the diagram, the process of transforming the original data is called **serialization**, and the reverse process of transforming it back is called **deserialization**.

Serialization has two primary use cases:

- Transferring the data between services, acting as a common *language* between them
- Encoding and decoding arbitrary data for storage, allowing you to store complex data structures as byte arrays or regular strings

In *Chapter 2*, while scaffolding our applications, we created our HTTP API endpoints and set them to return JSON responses to the callers. In that case, JSON played the role of a **serialization format**, allowing us to transform our data structures into it and then decode them back.

Let's take our `Metadata` structure defined in the `metadata/pkg/model/metadata.go` file as an example:

```go
// Metadata defines the movie metadata.
type Metadata struct {
    ID          string `json:"id"`
    Title       string `json:"title"`
    Description string `json:"description"`
    Director    string `json:"director"`
}
```

Our structure includes the records called **annotations** that help the JSON encoder transform our record into an output. For example, we create an instance of our structure:

```go
Metadata{
    ID:          "123",
    Title:       "The Movie 2",
    Description: "Sequel of the legendary The Movie",
    Director:    "Foo Bars",
}
```

When we then encode it with JSON, the result would be the following:

```json
{"id":"123","title":"The Movie 2","description":"Sequel of the legendary The Movie","director":"Foo Bars"}
```

Once the data is serialized, it can be used in many different ways. In our examples in *Chapter 2*, we used JSON format for sending and receiving the data between our microservices. Some additional use cases of serialization include the following:

- **Store configuration**: Serialization formats are commonly used for storing configuration. For example, you can define your service settings using these kinds of formats and then read them in your service code.

- **Store records in a database**: Formats such as JSON are frequently used for storing arbitrary data in databases. For example, key-value databases require encoding entire record values into byte arrays, and developers often use formats such as JSON to encode and decode these record values.

- **Logging**: Application logs are often stored in JSON format, making them easy to read for both humans and various applications, such as data visualization software.

JSON is one of the most popular serialization formats at the moment and it has been essential to web development. It has the following benefits:

- **Language support**: Most programming languages include tools for encoding and decoding JSON.

- **Browser support**: JSON has been an integral part of web applications and all modern browsers include developer tools to work with it in the browser itself.

- **Readability**: JSON records are easily readable and are often easy to use during both the development and debugging of web applications.

However, it has certain limitations as well:

- **Size**: JSON is not a size-efficient format. In this chapter, we are going to see which formats and protocols provide output records that are smaller in size.

- **Speed**: As with its output size, the encoding and decoding speed with JSON is not the fastest when set against other popular serialization protocols.

Let's explore the other popular serialization formats.

Popular serialization formats

There are many popular serialization formats and protocols used in the industry. Let's cover some of the most popular formats:

- **XML**
- **YAML**
- **Apache Thrift**
- **Apache Avro**
- **Protocol Buffers**

This section will provide a high-level overview of each one, as well as some key differences between these protocols.

XML

XML is one of the earliest serialization formats for web service development. It was created in 1998 and is still widely used in the industry, especially in enterprise applications.

XML represents data as a tree of nodes called elements. An element example would be `<example>Some value</example>`. If we serialized our metadata structure mentioned above, the result would be the following:

```
<Metadata><ID>123</ID><Title>The Movie 2</
Title><Description>Sequel of the legendary The Movie</
Description><Director>Foo Bars</Director></Metadata>
```

You may notice that the serialized XML representation of our data is slightly longer than the JSON one. It is one of the downsides of XML format – the output is often the largest among all popular serialization protocols, making it harder to read and transfer the data. On the other hand, XML's advantages include its wide adoption and popularity, readability, as well as its wide library support.

YAML

YAML is a serialization format that was first released in 2001. It gained popularity over the years, becoming one of the most popular serialization formats in the industry. Designers of the language took a strong focus on its readability and compactness, making it a perfect tool for defining arbitrary human-readable data. We can illustrate this on our Metadata structure: in the YAML format, it would look as follows:

```
metadata:
   id: 123
   title: The Movie 2
   description: Sequel of the legendary The Movie
   director: Foo Bars
```

The YAML format is widely used for storing configuration data. One of the reasons for this is the ability to include comments, which is lacking in other formats, such as JSON. The use of YAML for service-to-service communication is less common, primarily due to the greater size of the serialized data. Let's get to some more size-efficient serialization formats.

Apache Thrift

So far, we have reviewed JSON, XML, and YAML, and all are primarily used for defining and serializing arbitrary types of data. There are other solutions to a broader class of problems, when we want not only to serialize and deserialize the data but also to transfer it between multiple services. These solutions combine two roles: they act as both serialization formats and **communication protocols** — mechanisms for sending and receiving arbitrary data over the network. HTTP is an example of such a protocol but developers are not limited to using it in their applications.

Apache Thrift is a combination of serialization and a communication protocol that can be used for both defining your data types and allowing your services to communicate with each other by passing them. It was initially created at Facebook but later became a community-supported open source project under the Apache Software Foundation.

Thrift, unlike JSON and XML, requires you to define your structures in their own format first. In our example, for the Metadata structure, we would need to create a file with the `.thrift` extension, including the definition in Thrift language:

```
struct Metadata {
    1: string id,
    2: string title,
    3: string description,
    4: string director
}
```

Once you have a Thrift file, you can use it with an automatic Thrift code generator to generate the code for most programming languages that would contain the defined structures and logic to encode and decode it. In addition to data structures, Thrift allows you to define **Thrift services** — sets of functions that can be called remotely. Here's an example of a Thrift service definition:

```
service MetadataService {
    Metadata get(1: string id)
}
```

The example here defines a service called `MetadataService`, which provides a `get` function, returning a `Metadata` Thrift object. A Thrift-compatible server can act as such a Thrift service, processing incoming requests from the client applications — we are going to learn how to write such servers in *Chapter 5*.

Let's explore the benefits and limitations of Apache Thrift. The benefits include the following:

- A smaller output size and higher encoding and decoding speed compared to XML and JSON. Thrift-serialized data can be 30 to 50% smaller in size than XML and JSON.

- The ability to define not only structures but entire services and generate code for them, allowing communication between the servers and their clients.

The limitations include the following:

- Relatively low popularity and adoption in recent years due to moving to more popular and efficient formats.

- It lacks official documentation. Thrift is a relatively complex technology, and most documentation is unofficial.

- Unlike JSON and XML, Thrift-serialized data is not readable, so it's trickier to use it for debugging.

- Nearly no support in recent years – Facebook keeps maintaining a separate branch of it called Facebook Thrift, but it is much less popular than its Apache counterpart.

Let's see the other popular serialization formats that are widely used across the industry.

Apache Avro

Apache Avro is a combination of a serialization format and a communication protocol that is somewhat similar to Apache Thrift. Apache Avro also requires a developer to define a schema (written either in JSON or in its own language called Avro IDL) for their data. In our case, the Metadata structure would have the following schema:

```
{
    "namespace": "example.avro",
    "type": "record",
    "name": "Metadata",
    "fields": [
        {"name": "id", "type": "string"},
        {"name": "title", "type": "string"},
        {"name": "description", "type": "string"},
        {"name": "director", "type": "string"},
    ]
}
```

Then, the schema would be used for translating the structures into a serialized state and back.

It is not uncommon for types and structures to change over time, and microservice API and structure definitions need to evolve. With Avro, developers can create a new version of a schema (represented as a separate file, often suffixed with an incremental version number), and keep both the old and new versions in the code base. This way, the application can encode and decode data in either format, even if they have some incompatible changes, such as changes in field names. This is one of the key benefits of using Apache Avro over many other serialization protocols. Additionally, Apache Avro allows you to generate code for existing schemas, making it easier to translate between serialized data and corresponding data structures in different programming languages.

Protocol Buffers

Protocol Buffers is a serialization format that was created at Google more than 20 years ago. In 2008, the format became public and immediately gained popularity among developers. The benefits of the format include the following:

- The simplicity of the definition language
- A small data output size
- High performance of serialization and deserialization
- The ability to define services in addition to data structures and compile client and server code in multiple languages
- Protocol evolution and official support by Google

The popularity of Protocol Buffers and its simplicity, as well as the efficiency of its data encoding, makes it a great fit for using it in microservice development. We are going to use Protocol Buffers for serializing and deserializing the data transferred between our services, as well as defining our service APIs. In the next section, you will learn how to start using Protocol Buffers and move our microservice logic to Protocol Buffers from JSON.

Using Protocol Buffers

In this section, we are going to illustrate how you can use Protocol Buffers for your applications. We will use the microservice examples from the previous chapters and define our data model in the Protocol Buffers format. Then, we will be using the code generation tools with Protocol Buffers to generate our data structures. Finally, we will illustrate how to use our generated code for serializing and deserializing our data.

First, let's prepare our application. Create the directory called `api` under our application's `src` directory. Inside this directory, create a `movie.proto` file and add the following to it:

```
syntax = "proto3";
option go_package = "/gen";

message Metadata {
    string id = 1;
    string title = 2;
    string description = 3;
    string director = 4;
}
```

```
message MovieDetails {
    float rating = 1;
    Metadata metadata = 2;
}
```

Let's describe the code we just added. In the first line, we set the syntax to `proto3`, the latest version of the Protocol Buffers protocol. The second line defines the output path for the code generated. The rest of the file includes two structures that we need for our microservices, similar to the Go structures we created in *Chapter 2*.

Now, let's generate the code for our structures. In the `src` directory of our application, run the following command:

```
protoc -I=api --go_out=. movie.proto
```

If the command executes successfully, you should find a new directory called `src/gen`. The directory should include a file called `movie.pb.go` with the generated code that includes our structures and the code to serialize and deserialize them. For example, the generated `MovieDetails` structure code would be the following:

```
type Metadata struct {
    state         protoimpl.MessageState
    sizeCache     protoimpl.SizeCache
    unknownFields protoimpl.UnknownFields

    Id          string `protobuf:"bytes,1,opt,name=id,proto3"
json:"id,omitempty"`
    Title       string
`protobuf:"bytes,2,opt,name=title,proto3"
json:"title,omitempty"`
    Description string
`protobuf:"bytes,3,opt,name=description,proto3"
json:"description,omitempty"`
    Director    string
`protobuf:"bytes,4,opt,name=director,proto3"
json:"director,omitempty"`
}
```

Let's now describe what exactly we have just achieved. We have created a `movie.proto` file that defines our **data schema** — the definition of our data structures. The schema is now defined independently from our Go code, providing the following benefits to us:

- **Explicit schema definition**: Our data schema is now decoupled from the code and explicitly defines the application data types. This makes it easier to see the data types provided by application APIs.

- **Code generation**: Our schema can be converted to code via code generation. We are going to use it later in *Chapter 5* for sending the data between the services.

- **Cross-language support**: We can generate our code not only for Go but also for other programming languages. If our model changes, we would not need to rewrite our structures for all languages. Instead, we can just re-generate the code for all languages by running a single command.

Let's do a quick benchmark and compare the size of serialized data for three serialization protocols – XML, JSON, and Protocol Buffers. For this, let's write a small tool to do so.

Inside the `src` directory, create a directory called `cmd/sizecompare` and add a `main.go` file to it with the following contents:

```
package main

import (
    "encoding/json"
    "encoding/xml"
    "fmt"

    "github.com/golang/protobuf/proto"
    "movieexample.com/gen"
    "movieexample.com/metadata/pkg/model"
)

var metadata = &model.Metadata{
    ID:          "123",
    Title:       "The Movie 2",
    Description: "Sequel of the legendary The Movie",
    Director:    "Foo Bars",
}

var genMetadata = &gen.Metadata{
```

```
        Id:         "123",
        Title:      "The Movie 2",
        Description: "Sequel of the legendary The Movie",
        Director:   "Foo Bars",
    }
```

Let's implement the main function:

```
func main() {
    jsonBytes, err := serializeToJSON(metadata)
    if err != nil {
        panic(err)
    }

    xmlBytes, err := serializeToXML(metadata)
    if err != nil {
        panic(err)
    }

    protoBytes, err := serializeToProto(genMetadata)
    if err != nil {
        panic(err)
    }

    fmt.Printf("JSON size:\t%dB\n", len(jsonBytes))
    fmt.Printf("XML size:\t%dB\n", len(xmlBytes))
    fmt.Printf("Proto size:\t%dB\n", len(protoBytes))
}
```

Additionally, add the following functions:

```
func serializeToJSON(m *model.Metadata) ([]byte, error) {
    return json.Marshal(m)
}

func serializeToXML(m *model.Metadata) ([]byte, error) {
    return xml.Marshal(m)
}
```

```
func serializeToProto(m *gen.Metadata) ([]byte, error) {
    return proto.Marshal(m)
}
```

In the preceding code, we encode our `Metadata` structure using JSON, XML, and Protocol Buffers formats, and print the output sizes in bytes for each encoded result.

You may need to fetch the `github.com/golang/protobuf/proto` package required for our benchmark by running the following:

```
go mod tidy
```

Now, you can run our benchmark by executing `go run *.go` inside its directory and will see following the output:

```
JSON size: 106B
XML size: 148B
Proto size: 63B
```

The result is quite interesting. The XML output is almost 40% bigger than the JSON one. At the same time, Protocol Buffers's output is more than 40% smaller than the JSON data and more than twice as small as the XML result. This illustrates quite well how efficient the Protocol Buffers format is compared to the other two in terms of output size. By switching from JSON to Protocol Buffers, we reduce the amount of data that we need to send over the network and make our communication faster.

Let's now do an additional experiment and test serialization speed for all three formats. For this, we are going to do a **benchmark** — an automated performance check that is going to measure how fast a target operation is.

Create a file called `main_test.go` in the same directory and add the following to it:

```
package main

import (
    "testing"
)

func BenchmarkSerializeToJSON(b *testing.B) {
    for i := 0; i < b.N; i++ {
        serializeToJSON(metadata)
    }
```

```
}

func BenchmarkSerializeToXML(b *testing.B) {
    for i := 0; i < b.N; i++ {
        serializeToXML(metadata)
    }
}

func BenchmarkSerializeToProto(b *testing.B) {
    for i := 0; i < b.N; i++ {
        serializeToProto(genMetadata)
    }
}
```

We have just created a Go benchmark, that is going to tell us how fast is JSON, XML, and Protocol Buffers encoding. We are going to cover the details of benchmarking in *Chapter 8*, let's now run the code to see the output by executing the following command:

go test -bench=.

The result of the command should look as follows:

```
goos: darwin
goarch: amd64
pkg: movieexample.com/cmd/sizecompare
cpu: Intel(R) Core(TM) i7-8850H CPU @ 2.60GHz
BenchmarkSerializeToJSON-12            3308172                   342.2
ns/op
BenchmarkSerializeToXML-12              480728                2519 ns/
op
BenchmarkSerializeToProto-12           6596490                   185.7
ns/op
PASS
ok      movieexample.com/cmd/sizecompare     5.239s
```

You can see the names of three functions that we just implemented and two numbers next to them:

- The first one is the number of times the function got executed
- The second is the average processing speed, measured in nanoseconds per operation

From the output, we can see that Protocol Buffers serialization on average took 185.7 nanoseconds, while JSON serialization was almost two times slower at 342.2 nanoseconds. XML serialization on average took 2519 nanoseconds, being more than 13 times slower than Protocol Buffers, and more than 7 times slower than JSON serialization.

The benchmark is indeed interesting – it illustrates how different the average encoding speeds for various serialization formats are. If performance is important for your services, you should consider faster serialization formats to achieve a higher encoding and decoding speed.

For now, we are going to leave the generated structures in our repository. We will be using them in the next chapter, *Chapter 5*, to replace our JSON API handlers.

Now, let's learn some best practices for using serialization.

Best practices for serialization

This section summarizes the best practices for serializing and deserializing the data. These practices will help you make efficient decisions for using serialization in your applications and writing your schema definitions in Protocol Buffers and other formats:

- **Keeping your schema backward compatible**: Avoid any changes in your data schema that would break any existing callers. Such changes include modifications (renaming or removal) of field names and types.

- **Ensuring that data schemas are kept in sync between clients and servers**: For serialization formats with explicit schema definitions, such as Apache Thrift, Protocol Buffers, and Apache Avro, you should keep clients and servers in sync with the latest schema versions.

- **Documenting implicit details**: Let the callers know any implicit details related to your data schema. For example, if your API does not allow an empty value of a certain field of a structure, include this in the comments in the schema file.

- **Using built-in structures for representing time whenever possible**: Protocol Buffers and some other serialization protocols provide built-in types for timestamps and durations. Taking Protocol Buffers as an example, having a `int timestamp` filed would be considered a bad practice. The right approach would be to use `google.protobuf.Timestamp`.

- **Using consistent naming**: Opt for using consistent naming in your schema files, similarly to your code.

- **Following the official style guide**: Get familiar with the official style guide if you are using a schema definition language, such as Thrift or Protocol Buffers. You can find the link to the official style guide for Protocol Buffers in the *Further reading* section below.

This list provides some high-level recommendations applicable to all serialization protocols. For protocol-specific recommendations, follow the official documentation and check the popular open source projects to get some real-world code examples.

Summary

In this chapter, we covered the basics of serialization and illustrated how our data structures could be encoded using various serialization protocols, including XML, JSON, and Protocol Buffers. You learned about the differences between the most popular serialization protocols and their main advantages and disadvantages.

We covered the basics of Protocol Buffers and showed how to define custom data structures in its schema definition language. Then, we used the example code to illustrate how to generate the schema files for the Go language. Finally, we covered the differences in compression efficiency between XML, JSON, and Protocol Buffers.

In the next chapter, we are going to continue using Protocol Buffers and will show how to use it for communication between services.

Further reading

- *The Protocol Buffers documentation*: `https://developers.google.com/protocol-buffers`

- *Comparison of serialization formats*: `https://en.wikipedia.org/wiki/Comparison_of_data-serialization_formats`

- *The Protocol Buffers official style guide*: `https://developers.google.com/protocol-buffers/docs/style`

5

Synchronous Communication

In this chapter, we are going to cover the most common way of communicating between microservices – synchronous communication. In *Chapter 2*, we already implemented the logic for communicating between microservices via the HTTP protocol and returning results in JSON format. In *Chapter 4*, we illustrated that the JSON format is not the most efficient in terms of data size, and there are many different formats providing additional benefits to developers, including code generation. In this chapter, we are going to show you how to define service APIs using Protocol Buffers and generate both client and server code for them.

By the end of this chapter, you will understand the key concepts of synchronous communication between microservices and will have learned how to implement microservice clients and servers.

The knowledge you gain in this chapter will help you to learn how to better organize the client and server code, generate the code for serialization and communication, and use it in your microservices. In this chapter, we will cover the following topics:

- Introduction to synchronous communication
- Defining a service API using Protocol Buffers
- Implementing gateways and clients

Now, let's proceed to the main concepts of synchronous communication.

Technical requirements

To complete this chapter, you will need Go 1.11, a Protocol Buffers compiler that we installed in the previous chapter, and a gRPC plugin.

You can install the gRPC plugin by running the following command:

```
go install google.golang.org/grpc/cmd/protoc-gen-go-grpc@latest
export PATH="$PATH:$(go env GOPATH)/bin"
```

You can find the GitHub code for this chapter at `https://github.com/PacktPublishing/` `microservices-with-go/tree/main/Chapter05`.

Introduction to synchronous communication

In this section, we are going to cover the basics of synchronous communication and introduce you to some additional benefits of Protocol Buffers that we are going to use for our microservices.

Synchronous communication is the way of interaction between network applications, such as microservices, in which services exchange data using a **request-response model**. The process is illustrated in the following diagram:

Figure 5.1 – Synchronous communication

There are many **protocols** allowing applications to communicate in this way. HTTP is among the most popular protocols for synchronous communication. In *Chapter 2*, we already implemented the logic for calling and handling HTTP requests in our microservices.

The HTTP protocol allows you to send request and response data in different ways:

- **URL parameters**: In the case of the `https://www.google.com/search?q=portugal` URL, `q=portugal` is a URL parameter.

- **Headers**: Each request and response includes optional key-value pairs called headers, allowing you to propagate additional metadata, such as the client or browser name; for example, `User-Agent: Mozilla/5.0`.

- **Request and response body**: The request and response can include a body that contains arbitrary data. For example, when a client uploads a file to a server, the file contents are usually sent as a request body.

  When a server cannot handle a client request due to an error or the request is not received due to network issues, the client receives a specific response indicating an error. In the case of the HTTP protocol, there are two types of errors:

  - **Client error**: This error is caused by the client. Examples of such errors include invalid request arguments (such as an incorrect username), unauthorized access, and access to a resource that is not found (for example, a non-existing web page).

  - **Server error**: This error is caused by the server. This could be an application bug or an error with an upstream component, such as a database.

In *Chapter 2*, we implemented our API handlers by sending the result data as an HTTP response body in JSON format. We achieved this by using the Go JSON encoder:

```
if err := json.NewEncoder(w).Encode(details); err != nil {
    log.Printf("Response encode error: %v\n", err)
}
```

As discussed in the previous chapter, the JSON format is not the most optimal in terms of data size. Also, it does not offer useful tools, such as the cross-language code generation of data structures, that are provided by the formats, such as Protocol Buffers. Additionally, sending requests over HTTP and encoding the data manually is not the only form of communication between the services. There are some existing **remote procedure call** (**RPC**) libraries and frameworks that help to communicate between multiple services and offer some additional features to application developers:

- **Client and server code generation**: Developers can generate the client code for connecting and sending data to other microservices, as well as generate the server code for accepting incoming requests.

- **Authentification**: Most RPC libraries and frameworks offer authentication options for cross-service requests, such as TLS-based and token-based authentication.

- **Context propagation**: This is the ability to send additional data with requests, such as traces, which we are going to cover in *Chapter 11*.

- **Documentation generation**: Thrift can generate HTML documentation for services and data structures.

In the next section, we are going to cover some of the RPC libraries that you can use in your Go services, along with the features they provide.

Go RPC frameworks and libraries

Let's review some popular RPC frameworks and libraries that are available for Go developers.

Apache Thrift

We already covered Apache Thrift in *Chapter 4* and mentioned its ability to define RPC services – sets of functions provided by an application, such as a microservice. Here is an example of a Thrift RPC service definition:

```
service MetadataService {
  Metadata get(1: string id)
}
```

The Thrift definition of a service can be used to generate client and server code. The client code would include the logic for connecting to an instance of a service, as well as making requests to it, serializing and deserializing the request and response structures. The advantage of using a library such as Apache Thrift over making HTTP requests manually is the ability to generate such code for multiple languages: a service written in Go could easily talk to a service written in Java, while both would use the generated code for the communication, removing the need of implementing serialization/deserialization logic. Additionally, Thrift allows us to generate the documentation for RPC services.

gRPC

gRPC is an RPC framework that was created at Google. gRPC uses HTTP/2 as the transport protocol and Protocol Buffers as a serialization format. Similar to Apache Thrift, it provides an ability to define RPC services and generate the client and server code for the services. In addition to this, it offers some extra features, such as the following:

- Authentication
- Context propagation
- Documentation generation

gRPC adoption is much higher than for Apache Thrift, and its support of the popular Protocol Buffers format makes it a great fit for microservice developers. In this book, we are going to use gRPC as a framework for synchronous communication between our microservices. In the next section, we are going to illustrate how to leverage the features provided by Protocol Buffers to define our service APIs.

Defining a service API using Protocol Buffers

Let's demonstrate how to define a service API using the Protocol Buffers format and generate the client and server gRPC code for communication with each of our services using a **proto** compiler. This knowledge will help you to establish a foundation for both defining and implementing APIs for your microservices using one of the industry's most popular communication tools.

Let's start with our metadata service and write its API definition in the Protocol Buffers schema language.

Open the api/movie.proto file that we created in the previous chapter and add the following to it:

```
service MetadataService {
    rpc GetMetadata(GetMetadataRequest) returns
(GetMetadataResponse);
    rpc PutMetadata(PutMetadataRequest) returns
(PutMetadataResponse);
```

```
}

message GetMetadataRequest {
    string movie_id = 1;
}

message GetMetadataResponse {
    Metadata metadata = 1;
}
```

The code we just added defines our metadata service and its GetMetadata endpoint. We already have the Metadata structure from the previous chapter that we can reuse now.

Let's note some aspects of the code we just added:

- **Request and response structures**: It's a good practice to create a new structure for both a request and a response. In our example, they are GetMetadataRequest and GetMetadataResponse.

- **Naming**: You should follow consistent naming rules for all your endpoints. We are going to prefix all request and response functions with the function name.

Now, let's add the definition of the rating service to the same file:

```
service RatingService {
    rpc GetAggregatedRating(GetAggregatedRatingRequest) returns
(GetAggregatedRatingResponse);
    rpc PutRating(PutRatingRequest) returns
(PutRatingResponse);
}

message GetAggregatedRatingRequest {
    string record_id = 1;
    int32 record_type = 2;
}

message GetAggregatedRatingResponse {
    double rating_value = 1;
}
```

```
message PutRatingRequest {
    string user_id = 1;
    string record_id = 2;
    int32 record_type = 3;
    int32 rating_value = 4;
}


message PutRatingResponse {
}
```

Our rating service has two endpoints, and we defined requests and responses for them in a similar way to the metadata service.

Finally, let's add the definition of the movie service to the same file:

```
service MovieService {
    rpc GetMovieDetails(GetMovieDetailsRequest) returns
(GetMovieDetailsResponse);
}

message GetMovieDetailsRequest {
    string movie_id = 1;
}

message GetMovieDetailsResponse {
    MovieDetails movie_details = 1;
}
```

Now our movie.proto file includes both our structure definitions and the API definitions for our services. We are ready to generate code for the newly added service definitions. In the src directory of the application, run the following:

```
protoc -I=api --go_out=. --go-grpc_out=. movie.proto
```

The preceding command is similar to the command that we used in the previous chapter for generating code for our data structures. However, it also passes a --go-grpc_out flag to the compiler. This flag tells the Protocol Buffers compiler to generate the service code in gRPC format.

Let's see the compiled code that was generated as the output for our command. If the command is executed without any errors, you will find a `movie_grpc.pb.go` file inside the `src/gen` directory. The file will include the generated Go code for our services. Let's take a look at the generated client code:

```go
type MetadataServiceClient interface {
    GetMetadata(ctx context.Context, in *GetMetadataRequest,
opts ...grpc.CallOption) (*GetMetadataResponse, error)
}


type metadataServiceClient struct {
    cc grpc.ClientConnInterface
}


func NewMetadataServiceClient(cc grpc.ClientConnInterface)
MetadataServiceClient {
    return &metadataServiceClient{cc}
}


func (c *metadataServiceClient) GetMetadata(ctx context.
Context, in *GetMetadataRequest, opts ...grpc.CallOption)
(*GetMetadataResponse, error) {
    out := new(GetMetadataResponse)
    err := c.cc.Invoke(ctx, "/MetadataService/GetMetadata", in,
out, opts...)
    if err != nil {
        return nil, err
    }
    return out, nil
}
```

This generated code can be used in our applications to call our API from the Go applications. Additionally, we can generate such client code for other languages, such as Java, adding more arguments to the compiler command that we just executed. This is a great feature that can save us lots of time when writing microservice applications – instead of writing client logic for calling our services, we can use the generated clients and plug them into our applications.

In addition to the client code, the Protocol Buffers compiler also generates the service code that can be used for handling the requests. In the same `movie_grpc.pb.go` file, you will find the following:

```
type MetadataServiceServer interface {
    GetMetadata(context.Context, *GetMetadataRequest)
(*GetMetadataResponse, error)
    mustEmbedUnimplementedMetadataServiceServer()
}

func RegisterMetadataServiceServer(s grpc.ServiceRegistrar, srv
MetadataServiceServer) {
s.RegisterService(&MetadataService_ServiceDesc, srv)
}
```

We are going to use both the client and server code that we just saw in our application. In the next section, we are going to modify our API handlers to use the generated code and handle requests using the Protocol Buffers format.

Implementing gateways and clients

In this section, we are going to illustrate how to plug the generated client and server gRPC code into our microservices. This will help us to switch communication between them from JSON-serialized HTTP to Protocol Buffers gRPC calls.

Metadata service

In *Chapter 2*, we created our internal model structures, such as metadata, and in *Chapter 4*, we created their Protocol Buffers counterparts. Then, we generated the code for our Protocol Buffers definitions. As a result, we have two versions of our model structures – internal ones, as defined in `metadata/pkg/model`, and the generated ones, which are located in the `gen` package.

You might think that having two similar structures is now redundant. While there is certainly some level of redundancy in having such duplicate definitions, these structures practically serve different purposes:

- **Internal model**: The structures that you create manually for your application should be used across its code base, such as the repository, controller, and other logic.

- **Generated model**: Structures generated by tools such as the protoc compiler, which we used in the last two chapters, should only be used for serialization. The use cases include transferring the data between the services or storing the serialized data.

You might be curious why it's not recommended to use the generated structures across the application code base. There are multiple reasons for this, which are listed as follows:

- **Unnecessary coupling between the application and serialization format**: If you ever want to switch from one serialization format to another (for example, from Thrift to Protocol Buffers), and all your application code base uses generated structures for the previous serialization format, you would need to rewrite not only the serialization code but the entire application.

- **Generated code structure could vary between different versions**: While the field naming and high-level structure of the generated structures are generally stable between different versions of code generation tooling, the internal functions and structure of the generated code could vary from version to version. If any part of your application uses some generated functions that get changed, your application could break unexpectedly during a version update of a code generator.

- **Generated code is often harder to use**: In formats such as Protocol Buffers, all fields are always optional. In generated code, this results in lots of fields that can have nil values. For an application developer, this means doing more nil checks across all applications to prevent possible panics.

Because of these reasons, the best practice is to keep both internal structures and the generated ones and only use the generated structures for serialization. Let's illustrate how to achieve this.

We would need to add some **mapping** logic to translate the internal data structures and their generated counterparts. In the metadata/pkg/model directory, create a mapper.go file and add the following to it:

```go
package model

import (
    "movieexample.com/gen"
)


// MetadataToProto converts a Metadata struct into a
// generated proto counterpart.
func MetadataToProto(m *Metadata) *gen.Metadata {
    return &gen.Metadata{
        Id:          m.ID,
        Title:       m.Title,
        Description: m.Description,
        Director:    m.Director,
    }
}
```

```go
// MetadataFromProto converts a generated proto counterpart
// into a Metadata struct.
func MetadataFromProto(m *gen.Metadata) *Metadata {
    return &Metadata{
        ID:          m.Id,
        Title:       m.Title,
        Description: m.Description,
        Director:    m.Director,
    }
}
```

The code we just added transforms the internal model into the generated structures and back. In the following code block, we are going to use it in the server code.

Now, let's implement a gRPC handler for the metadata service that would handle the client requests to the service. In the `metadata/internal/handler` package, create a `grpc` directory and add a `grpc.go` file:

```go
package grpc

import (
    "context"
    "errors"

    "google.golang.org/grpc/codes"
    "google.golang.org/grpc/status"
    "movieexample.com/gen"
    "movieexample.com/metadata/internal/controller"
    "movieexample.com/metadata/internal/repository"
    "movieexample.com/metadata/pkg/model"
)

// Handler defines a movie metadata gRPC handler.
type Handler struct {
    gen.UnimplementedMetadataServiceServer
    svc *controller.MetadataService
}
```

```
// New creates a new movie metadata gRPC handler.
func New(ctrl *metadata.Controller) *Handler {
    return &Handler{ctrl: ctrl}
}
```

Let's implement the GetMetadataByID function:

```
// GetMetadataByID returns movie metadata by id.
func (h *Handler) GetMetadata(ctx context.Context, req *gen.
GetMetadataRequest) (*gen.GetMetadataResponse, error) {
    if req == nil || req.MovieId == "" {
        return nil, status.Errorf(codes.InvalidArgument, "nil
req or empty id")
    }
    m, err := h.svc.Get(ctx, req.MovieId)
    if err != nil && errors.Is(err, controller.ErrNotFound) {
        return nil, status.Errorf(codes.NotFound, err.Error())
    } else if err != nil {
        return nil, status.Errorf(codes.Internal, err.Error())
    }
    return &gen.GetMetadataResponse{Metadata: model.
MetadataToProto(m)}, nil
}
```

Let's highlight some parts of this implementation:

- The handler embeds the generated gen.UnimplementedMetadataServiceServer structure. This is required by a Protocol Buffers compiler to enforce future compatibility.
- Our handler implements the GetMetadata function in exactly the same format as defined in the generated MetadataServiceServer interface.
- We are using the MetadataToProto mapping function to transform our internal structures into the generated ones.

Now we are ready to update our main file and switch it to the gRPC handler. Update the metadata/cmd/main.go file, changing its contents to the following:

```
package main

import (
    "context"
```

```
    "flag"
    "fmt"
    "log"
    "net"
    "time"

    "google.golang.org/grpc"
    "movieexample.com/gen"
    "movieexample.com/metadata/internal/controller"
    grpchandler "movieexample.com/metadata/internal/handler/
grpc"
    "movieexample.com/metadata/internal/repository/memory"
    "movieexample.com/metadata/pkg/model"
)

func main() {
    log.Println("Starting the movie metadata service")
    repo := memory.New()
    svc := controller.New(repo)
    h := grpchandler.New(svc)
    lis, err := net.Listen("tcp", "localhost:8081")
    if err != nil {
        log.Fatalf("failed to listen: %v", err)
    }
    srv := grpc.NewServer()
    gen.RegisterMetadataServiceServer(srv, h)
    srv.Serve(lis)
}
```

The updated main function illustrates how we instantiate our gRPC server and start listening for requests in it. The rest of the function is similar to the one we had before.

We are done with the changes to the metadata service and can now proceed to the rating service.

Rating service

Let's create a gRPC handler for the rating service. In the `rating/internal/handler` package, create a `grpc` directory and add a `grpc.go` file with the following code:

```
package grpc

import (
    "context"
    "errors"

    "google.golang.org/grpc/codes"
    "google.golang.org/grpc/status"
    "movieexample.com/gen"
    "movieexample.com/rating/internal/controller"
    "movieexample.com/rating/pkg/model"
)

// Handler defines a gRPC rating API handler.
type Handler struct {
    gen.UnimplementedRatingServiceServer
    svc *controller.RatingService
}

// New creates a new movie metadata gRPC handler.
func New(svc *controller.RatingService) *Handler {
    return &Handler{ctrl: ctrl}
}
```

Now, let's implement the `GetAggregatedRating` endpoint:

```
// GetAggregatedRating returns the aggregated rating for a
// record.
func (h *Handler) GetAggregatedRating(ctx context.
Context, req *gen.GetAggregatedRatingRequest) (*gen.
GetAggregatedRatingResponse, error) {
    if req == nil || req.RecordId == "" || req.RecordType == ""
{
        return nil, status.Errorf(codes.InvalidArgument, "nil
```

```
req or empty id")
    }
    v, err := h.svc.GetAggregatedRating(ctx, model.
RecordID(req.RecordId), model.RecordType(req.RecordType))
    if err != nil && errors.Is(err, controller.ErrNotFound) {
        return nil, status.Errorf(codes.NotFound, err.Error())
    } else if err != nil {
        return nil, status.Errorf(codes.Internal, err.Error())
    }
    return &gen.GetAggregatedRatingResponse{RatingValue: v},
nil
}
```

Finally, let's implement the `PutRating` endpoint:

```
// PutRating writes a rating for a given record.
func (h *Handler) PutRating(ctx context.Context, req *gen.
PutRatingRequest) (*gen.PutRatingResponse, error) {
    if req == nil || req.RecordId == "" || req.UserId == "" {
        return nil, status.Errorf(codes.InvalidArgument, "nil
req or empty user id or record id")
    }
    if err := h.svc.PutRating(ctx, model.RecordID(req.
RecordId), model.RecordType(req.RecordType), &model.
Rating{UserID: model.UserID(req.UserId), Value: model.
RatingValue(req.RatingValue)}); err != nil {
        return nil, err
    }
    return &gen.PutRatingResponse{}, nil
}
```

Now, we are ready to update our `rating/cmd/main.go` file. Replace it with the following:

```
package main

import (
    "log"
    "net"
```

```
        "google.golang.org/grpc"
        "movieexample.com/gen"
        "movieexample.com/rating/internal/controller"
        grpchandler "movieexample.com/rating/internal/handler/grpc"
        "movieexample.com/rating/internal/repository/memory"
)

func main() {
    log.Println("Starting the rating service")
    repo := memory.New()
    svc := controller.New(repo)
    h := grpchandler.New(svc)
    lis, err := net.Listen("tcp", "localhost:8082")
    if err != nil {
        log.Fatalf("failed to listen: %v", err)
    }
    srv := grpc.NewServer()
    gen.RegisterRatingServiceServer(srv, h)
    srv.Serve(lis)
}
```

The way we start the service is similar to the metadata service. Now, we are ready to link the movie service to both the metadata and rating services.

Movie service

In the previous examples, we created gRPC servers to handle client requests. Now, let's illustrate how to add logic for calling our servers. This will help us to establish communication between our microservices via gRPC.

First, let's implement a function that we can reuse in our service gateways. Create the src/internal/ grpcutil directory, and add a file called grpcutil.go to it. Add the following code to it:

```
package grpcutil

import (
    "context"
    "math/rand"
    "pkg/discovery"
```

```
    "google.golang.org/grpc"
    "google.golang.org/grpc/credentials/insecure"
    "movieexample.com/pkg/discovery"
)


// ServiceConnection attempts to select a random service
// instance and returns a gRPC connection to it.
func ServiceConnection(ctx context.Context, serviceName string,
registry discovery.Registry) (*grpc.ClientConn, error) {
    addrs, err := registry.ServiceAddresses(ctx, serviceName)
    if err != nil {
        return nil, err
    }
    return grpc.Dial(addrs[rand.Intn(len(addrs))], grpc.
WithTransportCredentials(insecure.NewCredentials()))
}
```

The function that we just implemented will try to pick a random instance of the target service using the provided service registry, and then it will create a gRPC connection for it.

Now, let's create a gateway for our metadata service. In the movie/internal/gateway package, create a directory called metadata. Inside it, create a grpc directory with a metadata.go file, containing the following code:

```
package grpc

import (
    "context"

    "google.golang.org/grpc"
    "movieexample.com/gen"
    "movieexample.com/internal/grpcutil"
    "movieexample.com/metadata/pkg/model"
    "movieexample.com/pkg/discovery"
)
```

```go
// Gateway defines a movie metadata gRPC gateway.
type Gateway struct {
    registry discovery.Registry
}

// New creates a new gRPC gateway for a movie metadata
// service.
func New(registry discovery.Registry) *Gateway {
    return &Gateway{registry}
}
```

Let's implement the function for getting the metadata from a remote gRPC service:

```go
// Get returns movie metadata by a movie id.
func (g *Gateway) Get(ctx context.Context, id string) (*model.
Metadata, error) {
    conn, err := grpcutil.ServiceConnection(ctx, "metadata",
g.registry)
    if err != nil {
        return nil, err
    }
    defer conn.Close()
    client := gen.NewMetadataServiceClient(conn)
    resp, err := client.GetMetadataByID(ctx, &gen.
GetMetadataByIDRequest{MovieId: id})
    if err != nil {
        return nil, err
    }
    return model.MetadataFromProto(resp.Metadata), nil
}
```

Let's highlight some details of our gateway implementation:

- We use the grpcutil.ServiceConnection function to create a connection to our metadata service.

- We create a client using the generated client code from the gen package.

- We use the MetadataFromProto mapping function to convert the generated structures into internal ones.

Now we are ready to create a gateway for our rating service. Inside the `movie/internal/gateway` package, create a `rating/grpc` directory and add a `grpc.go` file with the following contents:

```go
package grpc

import (
    "context"
    "pkg/discovery"
    "rating/pkg/model"

    "google.golang.org/grpc"
    "movieexample.com/internal/grpcutil"
    "movieexample.com/gen"
)

// Gateway defines an gRPC gateway for a rating service.
type Gateway struct {
    registry discovery.Registry
}

// New creates a new gRPC gateway for a rating service.
func New(registry discovery.Registry) *Gateway {
    return &Gateway{registry}
}
```

Add the implementation of the `GetAggregatedRating` function:

```go
// GetAggregatedRating returns the aggregated rating for a
// record or ErrNotFound if there are no ratings for it.
func (g *Gateway) GetAggregatedRating(ctx context.Context,
recordID model.RecordID, recordType model.RecordType) (float64,
error) {
    conn, err := grpcutil.ServiceConnection(ctx, "rating",
g.registry)
    if err != nil {
        return 0, err
    }
    defer conn.Close()
```

```
    client := gen.NewRatingServiceClient(conn)
    resp, err := client.GetAggregatedRating(ctx, &gen.
GetAggregatedRatingRequest{RecordId: string(recordID),
RecordType: string(recordType)})
    if err != nil {
        return 0, err
    }
    return resp.RatingValue, nil
}
```

At this point, we are almost done with the changes. The last step is to update the `main` function of the movie service. Change it to the following:

```
package main

import (
    "context"
    "log"
    "net"

    "google.golang.org/grpc"
    "movieexample.com/gen"
    "movieexample.com/movie/internal/controller"
    metadatagateway "movieexample.com/movie/internal/gateway/
metadata/grpc"
    ratinggateway "movieexample.com/movie/internal/gateway/
rating/grpc"
    grpchandler "movieexample.com/movie/internal/handler/grpc"
"movieexample.com/pkg/discovery/static"
)

func main() {
    log.Println("Starting the movie service")
    registry := static.NewRegistry(map[string][]string{
        "metadata": {"localhost:8081"},
        "rating":   {"localhost:8082"},
        "movie":    {"localhost:8083"},
    })
```

```
    ctx := context.Background()
    if err := registry.Register(ctx, "movie",
"localhost:8083"); err != nil {
        panic(err)
    }
    defer registry.Deregister(ctx, "movie")
    metadataGateway := metadatagateway.New(registry)
    ratingGateway := ratinggateway.New(registry)
    svc := controller.New(ratingGateway, metadataGateway)
    h := grpchandler.New(svc)
    lis, err := net.Listen("tcp", "localhost:8083")
    if err != nil {
        log.Fatalf("failed to listen: %v", err)
    }
    srv := grpc.NewServer()
    gen.RegisterMovieServiceServer(srv, h)
    srv.Serve(lis)
}
```

You might have noticed that the format hasn't changed, and we just updated the imports for our gateways, changing them from HTTP to gRPC.

We are done with the changes to our services. Now the services can communicate with each other using the Protocol Buffers serialization, and you can run them using the go run *.go command inside each cmd directory.

Summary

In this chapter, we covered the basics of synchronous communication and learned how to make microservices communicate with each other using the Protocol Buffers format. We illustrated how to define our service APIs using the Protocol Buffers schema language and generate code that can be reused in microservice applications written in Go and other languages.

The knowledge you gained in this chapter should help you write and maintain the existing services using Protocol Buffers and gRPC. It also serves as an example of how to use code generation for your services. In the next chapter, we are going to continue our journey into different ways of communication by covering another model, asynchronous communication.

Further reading

- *gRPC*: https://grpc.io
- *HTTP/2 detailed overview*: https://web.dev/performance-http2

6

Asynchronous Communication

In the previous chapter, we illustrated how services can communicate with each other using a synchronous request-response model. There are other communication models that provide various benefits to the application developer, such as asynchronous communication, which we are going to cover in this chapter.

In this chapter, you are going to learn the basics of asynchronous communication and some common techniques for using it, as well as some benefits and challenges it brings to microservice developers. We will cover a popular piece of asynchronous communication software, Apache Kafka, and illustrate how to use it for establishing communication between our microservices.

In this chapter, we are going to cover the following topics:

- Asynchronous communication basics
- Using Apache Kafka for messaging
- Asynchronous communication best practices

Let's proceed to the basics of asynchronous communication.

Technical requirements

To complete this chapter, you need Go 1.11+ or above, similar to the previous chapters.

You can find the GitHub code for this chapter here: `https://github.com/PacktPublishing/microservices-with-go/tree/main/Chapter06`

Asynchronous communication basics

In this section, we are going to cover some theoretical aspects of asynchronous communication. You will learn the benefits and the common issues of an asynchronous communication model, and the common ways of using it, as well as getting some real-world examples of asynchronous communication.

Asynchronous communication is communication between a sender and one or multiple receivers, where a sender does not necessarily expect an immediate response to their messages. In the synchronous communication model, which we covered in *Chapter 5*, the caller sending the request would expect an immediate (or nearly immediate, taking into account network latency) response to it. In asynchronous communication, it may take an arbitrary amount of time for the receiver to respond to the request, or to not respond at all (for example, when receiving a no-reply notification).

We can illustrate the differences between the two models using two examples. An example of synchronous communication is a phone call – two people having a phone conversation are in direct and immediate communication with each other, and they expect to hear the responses in real time. An example of asynchronous communication is sending mail to people. It can take time to respond to such mail, and the sender does not expect an immediate response to their messages.

It does not mean, however, that asynchronous communication is necessarily slower than the synchronous model. In most cases, asynchronous processing is as quick as synchronous processing and often can be even faster: asynchronous processing is often much less interruptive and leads to higher processing efficiency. It's like replying to 10 emails, one by one, compared to switching between 10 parallel phone calls — the latter example of synchronous processing can sometimes be way slower due to context switching and frequent interruptions.

Benefits and challenges of asynchronous communication

The asynchronous communication model comes with its own benefits and challenges. Developers need to consider both in order to make a decision on whether to use this model. What are the benefits of using asynchronous communication? Let's find out:

- The first benefit is a more streamlined approach to processing messages. Imagine you have a service whose purpose is to process data and report the status to another service. The reporting service does not necessarily need to wait for any response back from the service it is reporting to, as it would do in the synchronous model. In asynchronous mode, it just needs confirmation that the status message was sent successfully. This is similar to sending a large number of postcards to your relatives – if you send a dozen postcards, you don't want to wait until each card gets delivered before sending the next one!

- The second benefit of an asynchronous communication model is the ability to de-couple the sending and processing of requests. Imagine a caller requesting to download a large video and a remote server that needs to perform such a task. In a synchronous model, the caller would be waiting in real time until the entire video is processed. This could easily take minutes and sometimes even hours, making such waiting very inefficient. Instead, such a task could be performed in an asynchronous way, where the caller would send the task to a server, get an acknowledgment that the task was received, and perform any other activity until an eventual notification of completion (or processing failure) is received.

- The third benefit of asynchronous communication is better load balancing. Certain applications can have uneven request loads and are prone to sudden spikes of requests. If communication is synchronous, the server needs to answer every request in real time, and this can easily overload it. Imagine a waiter in a restaurant receiving a thousand dinner orders – such a high number of requests would completely overwhelm the worker, and also affect the clients.

The benefits that we just described are quite significant, and in many cases, asynchronous communication is the only way to perform certain types of tasks or to provide better system performance. Examples of problems for which asynchronous communication is a good fit include the following:

- **Long-running processing tasks**: Long-running tasks, such as video processing, are often better done asynchronously. The caller requesting such processing would not necessarily need to wait until it is completed and would eventually get notified of the final result.

- **Send once, processed by multiple components**: Certain types of messages, such as status reports, can be processed by multiple independent components. Imagine a system where multiple employees need to receive the same message – instead of sending it to each one independently, the message can get published to a component that can be consumed by everyone interested.

- **High-performance sequential processing**: Certain types of operations are more efficient when performed sequentially and/or in batches. For example, some operations, such as writes to HDD are often more performant when done sequentially (an example would be writing a very large file sequentially, without any interruptions). For such scenarios, asynchronous processing offers great performance improvements compared to more interactive and interruptive synchronous communication because the receiver of such requests can control the processing speed and process tasks one after another.

While the described benefits of asynchronous communication may seem appealing, it is important to note that it often brings some difficult challenges to developers:

- **More complex error handling**: Imagine sending a message to your friend and not receiving a response back. Was it because the friend did not receive the message? Did something happen during this time? Did the response get lost? In synchronous communication, such as a phone call, we would immediately know if the friend is not available and would be able to call back. In the case of an asynchronous scenario, we would need to think about more possible issues, such as the ones we described.

- **Reliance on additional components for message delivery**: Certain asynchronous communication use cases, such as the publisher-subscriber or message broker models described in the next section, require additional software for delivering messages. Such software often performs additional operations, such as message batching and storing, bringing additional complexity to the system in exchange for additional features it provides.

- **Asynchronous data flow may seem non-intuitive to many developers and be more complex**: Unlike the synchronous request-response model, where each request is logically followed by a response to it, asynchronous communication may be **unidirectional** (no responses are received at all) or may require the caller to perform additional steps in order to receive a response (for example, when the response is sent as a separate notification). Because of this, data flow in asynchronous systems may be more complex than in synchronous request-response interactions.

Now, let's cover some asynchronous communication techniques and patterns that can help you in organizing your services and establishing asynchronous communication between them.

Techniques and patterns of asynchronous communication

There are various techniques that help to make the asynchronous interaction between multiple services more efficient in various scenarios, such as sending a message to multiple recipients. In this section, we are going to describe multiple patterns that help to facilitate such interactions.

Message broker

A **message broker** is an intermediary component in the communication chain that can play multiple roles:

- **Message delivery**: It performs the delivery of a message to one or multiple receivers.
- **Message transformation**: It transforms an incoming message into another format that can be later consumed by receivers.
- **Message aggregation**: It aggregates multiple messages into a single one for more efficient delivery or processing.
- **Message routing**: It routes incoming messages to the appropriate destination based on pre-defined rules.

When you send a postcard to your friend or a relative, the post office plays the role of a message broker, playing an intermediary role in delivering it to the destination. In this example, the main benefit of using the message broker would be the convenience of sending the message (postcard, in our example) without any need to think about how to deliver it. Another benefit of using message brokers is delivery guarantees. A message broker can provide various levels of guarantees for message delivery. Examples of these guarantees include the following:

- **At-least-once**: The message gets delivered at least once, but may be delivered multiple times in case of failures.
- **Exactly-once**: The message broker guarantees that the message gets delivered and it will be delivered exactly once.
- **At-most-once**: The message can be delivered 0 or 1 time.

The exactly-once guarantee is often harder to achieve in practice than at-least-once and at-most-once. In the at-least-once model, a message broker can just re-send the message in case of any failure (such as a sudden power loss or a restart). In the exactly-once model, the message broker needs to perform additional checks or store extra metadata to ensure the message is never re-sent to the receiver in any possible case.

Another classification of message brokers is based on the possibility of them losing messages:

- **Lossy**: A message broker that can occasionally (for example, in case of failures) lose messages

- **Lossless**: A message broker that provides a guarantee of not losing any messages

The at-most-once guarantee is an example of a lossy message broker, and at-least-once and exactly-once brokers are examples of lossless ones. Lossy message brokers are faster than lossless ones because they don't need to handle extra logic for guaranteeing message delivery, such as persisting messages.

The publisher-subscriber model

The **publisher-subscriber model** is a model of communication between multiple components (such as microservices) where every component can publish messages and subscribe to the relevant ones.

Let's take Twitter as an example. Any user can publish messages to their feeds, and other users can subscribe to them. Similarly, microservices can communicate by publishing the data that other services can consume. Imagine that we have a set of services that process various types of user data, such as user photos, videos, and text messages. If a user deleted their profile, we would need to notify all services about this. Instead of notifying each service one by one, we could publish a single event that would indicate that a user profile is deleted, and all services could consume it and perform any relevant actions, such as archiving user data.

The relationship between the publishers, the subscribers, and the data produced by the publisher is illustrated in the following diagram:

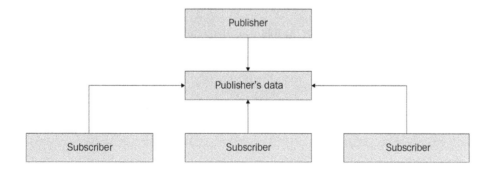

Figure 6.1 – The publisher-subscriber model

The publisher-subscriber model provides a flexible solution for sending and delivering data in a system where messages can be processed by multiple components. Each publisher can publish their messages without caring about the delivery process and any difficulties in delivering messages to an arbitrary number (even a very large one) of receivers. Each subscriber can subscribe to the relevant messages and get them delivered without needing to contact the publisher directly and check if there is any new data to consume. The latter feature is especially useful for scenarios with low message rates, such as occasional notification delivery.

Now, as we have covered some high-level asynchronous communication models, let's move on to the practical side of the chapter and illustrate how you can implement asynchronous communication in your microservices.

Using Apache Kafka for messaging

In this section, we are going to introduce you to Apache Kafka, a popular message broker system that we are going to use to establish asynchronous communication between our microservices. You will learn the basics of Kafka, how to publish messages to it, and how to consume such messages from the microservices we created in the previous chapters.

Apache Kafka basics

Apache Kafka is an open source message broker system that provides the ability to publish and subscribe to messages containing arbitrary data. Originally developed at LinkedIn, Kafka has become perhaps the most popular open source message broker software and is used by thousands of companies around the world.

In the Kafka model, a component that publishes messages is called a **producer**. Messages are published in sequential order to objects called **topics**. Each message in a topic has a unique numerical **offset** in it. Kafka provides APIs for consuming messages (the component for consuming messages is called a **consumer**) for the existing topics. Topics can also be partitioned to allow multiple consumers to consume from them (for example, for parallel data processing).

We can illustrate the Kafka data model in the following diagram:

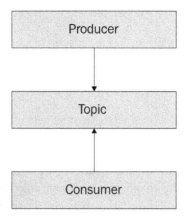

Figure 6.2 – The Apache Kafka data model

Having such a seemingly simple data model, Kafka is a powerful system that offers lots of benefits to its users:

- **High write and read throughput**: Kafka is optimized for highly performant write and read operations. It achieves this by doing as many sequential writes and reads as possible, allowing it to efficiently make use of hardware such as hard disk drives, as well as sequentially sending large amounts of data over the network.

- **Scalability**: Developers can leverage topic partitioning provided by Kafka to achieve more performant parallel processing of their data.

- **Flexible durability**: Kafka allows users to configure the policies for storing data, such as message retention. Messages can be stored for a fixed amount of time (for example, for 7 days) or indefinitely until there is enough space on the data storage.

> **Note**
> While Kafka provides many benefits for developers, it is important to note that it is a fairly complex infrastructure component that may be nontrivial to manage and maintain. We are going to use it in this chapter for illustrative purposes, especially taking into account its wide adoption and popularity in the developer community. In this chapter, we will avoid the difficulties of setting up a Kafka cluster by using its Docker version, but for production use cases you may need to get familiar with the relevant Kafka maintenance documentation, available at `https://kafka.apache.org/documentation/`

Let's explore how we can leverage the benefits offered by Kafka for the microservices we developed in the previous chapters.

Adopting Kafka for our microservices

Let's get back to the rating service example from the previous chapters. The service provided a synchronous API for inserting rating records, allowing its callers to call an endpoint and get an immediate response from the service. Such an API would be useful in many practical use cases, including one where the user submits a rating from a user interface or a web form.

Now consider a scenario where we work with a data provider who frequently publishes rating records (for example, movie ratings from a popular movie database, such as IMDb) that we can use in our rating service. Here, we would need to consume such records and ingest them into our system, so we could use them in addition to the data that was created through our API. The publisher-subscriber model that we described earlier in this chapter would be a great fit for this use case – the publisher would be the data provider that provides the rating data, and the subscriber would be a part of our application (such as a rating service), which would consume the data.

We can illustrate the described model using the following diagram:

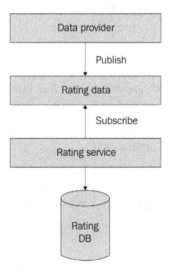

Figure 6.3 – The publisher-subscriber model of rating ingestion from a data provider

The model of interaction between the data provider and the rating service is a perfect example of asynchronous communication – the rating service does not necessarily need to process the provider's data immediately. It is up to us when and how to consume this data – our rating service could do this periodically (for example, once an hour, or once a day), or handle the new rating data as soon as it gets published. Let's choose the second approach in this chapter.

The only missing piece in our model is the component that allows us to publish the rating data from the data provider and subscribe to it from our rating service. Apache Kafka, which we described earlier, is a great fit for this use case – it provides a performant, scalable, and durable solution for producing and consuming arbitrary data, allowing us to use it as a rating data message broker.

To illustrate the model that we have just described, let's implement the following logic:

- A new example application that will produce rating data for Apache Kafka

- Logic in the rating service to consume the rating data from Apache Kafka and save it to our rating database

Before we proceed to implement both components, we need to decide which data serialization format to use between them. For simplicity, let's assume the data provider provides us with the rating data in JSON format. An example of the provided rating data would be as follows:

```
[{"userId":"105","recordId":"1","recordType":1,"value":5,"pr
oviderId":"test-provider","eventType":"put"},{"userId":"105"
,"recordId":"2","recordType":1,"value":4,"providerId":"test-
provider","eventType":"put"}]
```

Let's define a Go structure for such rating records. In the `src/rating/pkg/model/rating.go` file, add the following code:

```
// RatingEvent defines an event containing rating information.
type RatingEvent struct {
    UserID      UserID          `json:"userId"`
    RecordID    RecordID        `json:"recordId"`
    RecordType  RecordType      `json:"recordType"`
    Value       RatingValue     `json:"value"`
    EventType   RatingEventType `json:"eventType"`
}

// RatingEventType defines the type of a rating event.
type RatingEventType string

// Rating event types.
const (
    RatingEventTypePut    = "put"
    RatingEventTypeDelete = "delete"
)
```

Now, let's implement the example application that reads rating data from a provided file and produces it in Kafka. Create a cmd/ratingingester directory and add a main.go file, containing the following code:

```go
package main

import (
    "encoding/json"
    "fmt"
    "os"
    "time"

    "github.com/confluentinc/confluent-kafka-go/kafka"
    "movieexample.com/rating/pkg/model"
)

func main() {
    fmt.Println("Creating a Kafka producer")

    producer, err := kafka.NewProducer(&kafka.
ConfigMap{"bootstrap.servers": "localhost"})
    if err != nil {
        panic(err)
    }
    defer producer.Close()

    const fileName = "ratingsdata.json"
    fmt.Println("Reading rating events from file " + fileName)

    ratingEvents, err := readRatingEvents(fileName)
    if err != nil {
        panic(err)
    }

    const topic = "ratings"
    if err := produceRatingEvents(topic, producer,
ratingEvents); err != nil {
```

```
        panic(err)
    }


    const timeout = 10 * time.Second
    fmt.Println("Waiting " + timeout.String() + " until all
events get produced")


    producer.Flush(int(timeout.Milliseconds()))
}
```

In the code that we just added, we initialize a Kafka producer by calling `kafka.NewProducer`, read the rating data from a file, and produce rating events containing the rating data in Kafka. Note that we import the `github.com/confluentinc/confluent-kafka-go` Kafka library — a Kafka client made by Confluent, a company founded by the creators of Kafka. There are multiple popular open source Kafka libraries for Go, including `github.com/Shopify/sarama`, which is well maintained and is widely used across many Go projects. You can use either library in your projects depending on your preference.

Now, let's add a function for reading rating events to the file we just created:

```
func readRatingEvents(fileName string) ([]model.RatingEvent,
error) {
    f, err := os.Open(fileName)
    if err != nil {
        return nil, err
    }
    defer f.Close()
    var ratings []model.RatingEvent
    if err := json.NewDecoder(f).Decode(&ratings); err != nil {
        return nil, err
    }
    return ratings, nil
}
```

Finally, add a function for producing rating events:

```
func produceRatingEvents(topic string, producer kafka.Producer,
events []model.RatingEvent) error {
    for _, ratingEvent := range ratingEvents {
        encodedEvent, err := json.Marshal(ratingEvent)
```

```
    if err != nil {
        return err
    }


    if err := p.Produce(&kafka.Message{
TopicPartition: kafka.TopicPartition{Topic: &topic, Partition:
kafka.PartitionAny},
Value:            []byte(encodedEvent),
}, nil); err != nil {
        return err
    }
    return nil
}
```

Let's describe some parts of the code that we just wrote:

- We created a Kafka producer by calling a `kafka.NewProducer` function and providing `localhost` as the Kafka address for testing it locally.

- The program that we created is expected to read rating data from the `ratingsdata.json` file.

- When we produce events to Kafka using a `Produce` function, we specify a topic partition using a `kafka.TopicPartition` structure. In the structure, we provide the topic name (in our example, we call it `ratings`) and the topic partition (in our example, we use `kafka.PartitionAny` to produce a partition — we will cover this part later, in the *Asynchronous communication best practices* section).

- At the end of our main function, we call the `Flush` function to make sure all messages are sent to Kafka.

The function that we just created is using the `github.com/confluentinc/confluent-kafka-go/kafka` library, which we need to include in our Go module. Let's do this by running the following code:

```
go mod tidy
```

Let's also add a file containing the rating events. In the directory that we just used, create a `ratingsdata.json` file, containing the following code:

```
[{"userId":"105","recordId":"1","recordType":1,"value":5,"pr
oviderId":"test-provider","eventType":"put"},{"userId":"105"
,"recordId":"2","recordType":1,"value":4,"providerId":"test-
provider","eventType":"put"}]
```

Now, our application is ready. We have implemented the logic to read the rating data from a file and publish it to Apache Kafka for further consumption by the rating service. Let's implement the logic in the rating service to consume the published data. Create a `rating/internal/ingester/kafka` directory and add an `ingester.go` file with the following contents:

```go
package kafka

import (
    "context"
    "encoding/json"
    "fmt"
    "rating/pkg/model"

    "github.com/confluentinc/confluent-kafka-go/kafka"
    "movieexample.com/rating/pkg/model"
)

// Ingester defines a Kafka ingester.
type Ingester struct {
    consumer kafka.Consumer
    topic    string
}

// NewIngester creates a new Kafka ingester.
func NewIngester(addr string, groupID string, topic string)
(*Ingester, error) {
    consumer, err := kafka.NewConsumer(&kafka.ConfigMap{
        "bootstrap.servers": addr,
        "group.id":          groupID,
        "auto.offset.reset": "earliest",
    })
    if err != nil {
        return nil, err
    }
    return &Ingester{consumer, topic}, nil
}
```

Additionally, add this piece of code to it:

```go
// Ingest starts ingestion from Kafka and returns a channel
// containing rating events
// representing the data consumed from the topic.
func (i *Ingester) Ingest(ctx context.Context) (chan model.
RatingEvent, error) {
    if err := i.consumer.SubscribeTopics([]string{i.topic},
nil); err != nil {
        return nil, err
    }

    ch := make(chan model.RatingEvent, 1)
    go func() {
        for {
            select {
            case <-ctx.Done():
                close(ch)
                i.consumer.Close()
            default:
            }
            msg, err := i.consumer.ReadMessage(-1)
            if err != nil {
                fmt.Println("Consumer error: " + err.Error())
                continue
            }
            var event model.RatingEvent
            if err := json.Unmarshal(msg.Value, &event); err != nil
{
                fmt.Println("Unmarshal error: " + err.Error())
                continue
            }
            ch <- event
        }
    }()
    return ch, nil
}
```

In the code we just created, we have implemented a `NewIngester` function to create a new Kafka ingester, the component that will ingest rating events from it. The `Ingest` function starts message ingestion in the background and returns a Go channel with `RatingEvent` structures.

You may notice that in our call to the `ReadMessage` function, we provided `-1` as an argument. We specified a **consumer offset** — a checkpoint from which we should consume the messages from our topic. The value of `-1` is specific to Kafka and means that we will always consume from the beginning of the topic, reading all existing messages.

Let's use this structure in our rating service controller. In our `rating/internal/controller/controller.go` file, add the following code:

```go
type ratingIngester interface {
    Ingest(ctx context.Context) (chan model.RatingEvent, error)
}


// StartIngestion starts the ingestion of rating events.
func (s *RatingService) StartIngestion(ctx context.Context)
error {
    ch, err := s.ingester.Ingest(ctx)
    if err != nil {
        return err
    }
    for e := range ch {
        if err := s.PutRating(ctx, e.RecordID, e.RecordType,
&model.Rating{UserID: e.UserID, Value: e.Value}); err != nil {
            return err
        }
    }
    return nil
}
```

In our code, we call the `Ingest` function and get back a Go channel containing rating events from the topic. We iterate over it using the `for` operator. It keeps returning us available rating events until the channel is closed (for example, when the Kafka client is closed on service shutdown).

Now, update the existing `RatingService` structure and the New function in this file to the following:

```go
// RatingService encapsulates the rating service business
// logic.
type RatingService struct {
```

```
    repo      ratingRepository
    ingester ratingIngester
}

// New creates a rating service.
func New(repo ratingRepository, ingester ratingIngester)
*RatingService {
    return &RatingService{repo, ingester}
}
```

Now, our rating service is able to asynchronously consume rating events from Kafka, and execute the Put function for each one, writing it to the rating database. At this point, the rating service provides both a synchronous API for the callers that want to create ratings in real time and asynchronous logic for ingesting rating events from Apache Kafka.

We have covered the basics of asynchronous communication and illustrated how to use it in our microservices. Let's proceed to the final part of the chapter to see some best practices you should keep in mind while using this model.

Asynchronous communication best practices

In this section, we are going to cover the best practices of using the asynchronous communication model. You will learn some high-level recommendations for adopting the model in your applications and using it in a way that would maximize its benefits for you.

Versioning

Versioning is the technique of associating the format (or a schema) of the data with its version. Imagine you are working on a rating service, and you use a publisher-subscriber model for producing and consuming rating events. If at some point the format of your rating events gets changed, some of the events that are already produced will have an old data format, and some will have the new one. This situation may be hard to handle because the logic consuming such data would need to know how to differentiate between such formats and how to handle each one. Differentiating between two formats without knowing the data schema or its version could be a nontrivial task. Imagine that we have two JSON events:

```
{"recordID": "1", "rating": 5}
{"recordID": "2", "rating": 17, "userId": "alex"}
```

The second event has a `userId` field that is not present in the first. Is it because the producer did not provide it or because the data format did not have this field before?

Providing the schema version explicitly would help the data consumer handle this problem. Consider these updated examples:

```
{"recordID": "1", "rating": 5, "version": 1}
{"recordID": "2", "rating": 17, "userId": "alex", "version": 2}
```

In these examples, we know the versions of events and can now handle each one separately. For example, we may completely ignore events of a certain version (assume there was an application bug and we want to re-process events with an updated version instead) or use the version-specific validation (for instance, allow the records without a `userId` field for version 1, but disallow for the higher versions).

Versioning is very important to systems that can evolve over time because it makes dealing with different data formats easier. Even if you don't expect your data format to change, consider using versioning to increase your system's maintainability in the future.

Leveraging partitioning

In the code examples in the *Adopting Apache Kafka for our microservices* section, we implemented the logic for producing our data to message topics in Apache Kafka. The function for producing a message was as follows:

```
if err := p.Produce(&kafka.Message{
TopicPartition: kafka.TopicPartition{Topic: &topic, Partition:
kafka.PartitionAny},
Value:              []byte(encodedEvent),
}, nil); err != nil {
    return err
}
```

In this function, we used the `kafka.PartitionAny` option. As we mentioned in the *Apache Kafka basics* section, Kafka topics can be partitioned to allow multiple consumers to consume different partitions of a topic. Imagine you have a topic with three partitions – you can consume each one independently, as illustrated in the following diagram:

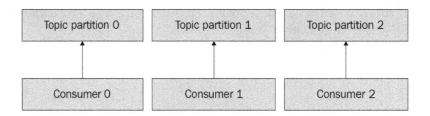

Figure 6.4 – A partitioned topic consumption example

You can control the number of topic partitions, as well as the partition for each message your services produce. Setting a partition manually may help you to achieve **data locality** — the ability to co-locate the data for various records, storing it together (in our use case, in the same topic partition). For example, you can partition the data using a user identifier, making sure the data for any user is stored on a single topic partition, helping you simplify the data search across the topic partitions.

The list of best practices that we just described is not comprehensive. It does not cover all recommendations for using asynchronous communication in your microservices, but it provides some great ideas for what you should consider. Get familiar with the articles listed in the *Further reading* section for some additional ideas and recommendations.

Summary

In this chapter, we have covered the basics of asynchronous communication and illustrated how to use it in your microservices. You have learned the benefits of asynchronous communication and the common patterns, such as publisher-subscriber and message broker. In addition to this, we have covered the basics of the Apache Kafka message broker and illustrated how to use it in our microservices and how to implement the logic for producing and consuming data from it.

In the next chapter, we are going to cover another important topic of microservice development – data storage. You will learn how to persist and read different types of service data, as well as how to implement the logic for working with MySQL database in your Go microservices.

Further reading

- *Apache Kafka documentation*: https://kafka.apache.org/documentation/
- *Publish-subscribe pattern*: https://en.wikipedia.org/wiki/Publish%E2%80%93subscribe_pattern
- *Asynchronous message-based communication*: https://docs.microsoft.com/en-us/dotnet/architecture/microservices/architect-microservice-container-applications/asynchronous-message-based-communication

7

Storing Service Data

In this chapter, we are going to review a very important topic: storing service data in persistent databases. In the previous chapters, we stored movie metadata and user ratings using in-memory repositories. While it was easy to implement in-memory storages of our data, using them would be impractical due to many reasons. One such reason is a lack of a persistence guarantee: if our service instances storing the data were restarted (for example, due to application failure or on host restart), we would lose all our data that was stored in the memory of a previously running instance. To guarantee that our data won't be lost over time, we need a solution that can persist our data and allow us to read and write it to our microservices. Among such solutions are databases, which we are going to review in this chapter.

We will cover the following topics:

- Introduction to databases
- Using MySQL to store our service data

Let's proceed to the first section of this chapter, which will provide an overview of microservice persistent storage solutions.

Technical requirements

To complete this chapter, you will need Go 1.11+ or above. Additionally, you will need the following tools:

- **Docker**: `https://www.docker.com`
- **grpcurl**: `https://github.com/fullstorydev/grpcurl`

You can find the GitHub code for this chapter here: `https://github.com/PacktPublishing/microservices-with-go/tree/main/Chapter07`.

Introduction to databases

Databases are systems that allow us to store and retrieve different types of data. Databases offer a variety of guarantees related to data storage, such as **durability** – a guarantee that all records and any related data changes will be persistent over time. A durability guarantee helps ensure that the data stored in a database won't be lost in case of various events such as software and hardware restarts, which are pretty common for microservices

Databases help solve lots of different other problems related to data storage. Let's illustrate one such problem using the metadata service that we created in *Chapter 2*. In our metadata service code, we implemented an in-memory repository for storing and retrieving the movie data that provides two functions, `Get` and `Put`. If we have just one instance of the metadata service, all its callers would be able to successfully write and read metadata records from the service memory, so long as a service instance does not perform a restart. However, let's imagine that we add another instance of a metadata service, as illustrated in the following diagram:

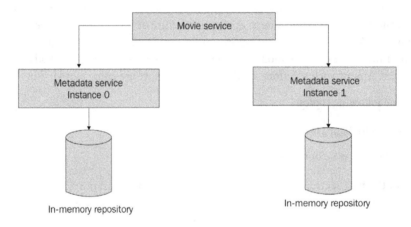

Figure 7.1 – Interaction between the movie service and two instances of the metadata service

Imagine that the movie service wants to write movie metadata and calls the metadata service to do this. The movie service instance would pick an instance of a metadata service (let's assume it picks instance 0) and send a write request to it, storing the record in memory of the instance that processes the request.

Now, let's assume the movie service wants to read the previously stored movie metadata and sends a read request to the metadata service. Depending on which instance handles the request, there would be two possible outcomes:

- **Instance 0**: Successfully return the previously saved movie metadata
- **Instance 1**: Return `ErrNotFound`

We just illustrated a case where the data is inconsistent between the two instances of the metadata repository. Because we have not implemented any coordination between our in-memory metadata repositories, each one acts as an independent data storage. Using our metadata service in such a way would be highly impractical: each newly added service instance would store a completely independent dataset.

To solve our data inconsistency problem, we can use a database to store the movie metadata: a database will handle all writes and reads from all the available metadata service instances, helping to co-locate the data from them inside a single type of logical storage. Such a scenario is illustrated in the following diagram:

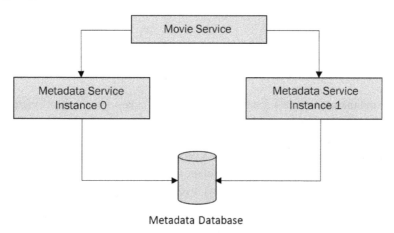

Figure 7.2 – Using a shared metadata database among multiple instances

In our diagram, multiple instances of a metadata service are using a shared database. This helps us aggregate and store the data coming from different metadata service instances.

We just illustrated how a database can help provide data durability to our services. There are other benefits that this can provide:

- **Transaction support**: Many databases support transactions – types of data changes – that have the following properties, abbreviated as **ACID**:

  - **Atomicity**: A change either happens entirely or does not happen at all

  - **Consistency**: A change brings a database from one valid state to another

  - **Isolation**: Concurrent changes get executed as they were executed sequentially

  - **Durability**: All changes get persisted

These transaction properties help provide a reliable way to modify different types of data (for example, simultaneously updating two financial account balances).

- **Data replication**: A database can offer data replication – a mechanism of duplicating data to additional instances, called **replicas**. Replication can help make the database more resilient to data losses (for example, when a database host becomes unavailable, data can be accessed on its replica) and reduce read latency (for example, when a user reads data from a replica that is located closer to them).

- **Additional query capabilities**: There are many databases (such as MySQL and PostgreSQL), that offer support for different query languages, such as SQL (`https://en.wikipedia.org/wiki/SQL`).

Different types of databases can help you efficiently store and retrieve various types of data. Let's review some of these popular database types:

- **Key-value databases**: These are databases that store data in key-value format, where each record contains a key (for example, user identifier) and a value (for example, user metadata). Keys and values are often represented as strings or byte arrays. Operations provided by key-value databases are often limited to key-based writes (store a value for the provided key) and key-based reads (read a value for the provided key). Because of their functional simplicity, key-value databases are among the most performant, since they don't involve any complex data processing or indexing.

- **Relational databases**: These are databases that store the data as a collection of tables, each consisting of a set of rows and columns. Users can run SQL queries to retrieve the data from one or multiple tables, being able to join the data between them or perform complex searching based on a variety of conditions. Historically, relational databases have been the most popular across all database types due to their ability to execute queries of any complexity, as well as their ability to store various types of structured (having a well-defined schema that maps row data to table columns) data.

- **Document databases**: These are databases that store data in a document format, such as JSON or XML. Document databases don't require you to define data schemas, so they're a great fit for storing various kinds of differently structured documents (for example, a collection of YAML files containing movie metadata from multiple websites in different formats).

- **Graph databases**: These are databases that store information in the form of *vertices* – objects that have different properties (for example, user details), and *edges* – the relationships between vertices (for example, if user A is following user B). Graph databases are different from relational databases in terms of the types of read queries they offer: most graph databases support *traversal* queries (checking each vertex in a graph), *connectivity* queries (getting all vertices connected to the target one), and many other ones.

- **Blob databases**: These are databases for storing *blob* (binary large object) data, such as audio or video files. Blob records are generally immutable (their content never gets changed after a successful write), so blob databases are well optimized for *append-only* writes (such as writing new files), as well as blob reads (such as retrieving the contents of a large file).

We won't go into the details of various types of databases because this is a topic for a separate book, but it is important to note that there is no *one-size-fits-all* solution among all of them, and each database provides a unique set of features that can be useful for solving a specific problem. For example, if the only purpose of your service is to store files, using a blob database would be sufficient, while a graph database could help with building a social network to store user relationship data. However, many use cases can be modeled using the relational model that powers relational databases: since it was introduced in 1970 by E.F. Codd, it has been used across the software development industry for nearly all types of problems. Popular relational databases, such as MySQL and PostgreSQL, remain the top-used software solutions, helping to build various types of applications, from tiny services running on a single host to large-scale clusters spanning hundreds of thousands of hosts. Because of the wide adoption and maturity of popular relational databases, many companies use them as a standard way to store various types of data. We are going to illustrate how to store our microservice data using a popular relational database: MySQL.

Using MySQL to store our service data

In this section, we are going to provide a brief overview of MySQL and demonstrate how to write and read data from the microservices that we created in the previous chapters.

MySQL is an open source relational database that was created in 1995 and since then has become one of the top-used databases across the development industry, according to DB-Engines ranking (`https://db-engines.com/en/ranking`). It stores data as a set of tables, each consisting of rows and columns of predefined types (such as string, numeric, binary, or more), and allows to data via SQL queries. For example, assume you have the following data, stored as a table called **movies**:

id	title	director
922	New York Stories	John Jones
1055	Christmas Day	Ben Miles
1057	Sunny Weather 3	Ben Miles

Table 7.1 – Movie table example

A SQL query to get all movies filmed by a particular director would look as follows:

```
SELECT * FROM movies WHERE director = "Ben Miles"
```

Now, let's imagine we have in the same database called **ratings** that contains the following data:

record_id	record_type	user_id	rating
1055	movie	alex001	5
1055	movie	chris.rocks	3
1057	movie	alex001	4

Table 7.2 – Rating table example

Using SQL language, we can write a more complex query to get all ratings that are associated with movies of a certain director:

```
SELECT * FROM ratings r INNER JOIN movies m ON r.record_id =
m.id WHERE r.record_type = "movie" AND m.director = "Ben Miles"
```

In our SQL query, we perform a **join** of two tables – an operation that allows us to group the data belonging to two tables and perform additional filtering (in our case, we will only select rating events where the **record_type** column's value is equal to *movie* and the **director** column's value is equal to *Ben Miles*).

To demonstrate how to use MySQL to store our service data, let's define which data we want to store and how we want to access it. Let's start with the metadata service, which performs two data storage operations:

- Stores movie metadata for a given movie ID
- Gets movie metadata for a given movie ID

Now, let's review the data schema of the movie metadata object. Our movie metadata contains the following fields:

- **ID**: String
- **Title**: String
- **Description**: String
- **Director**: String

Now, let's see which data is stored by our rating service, which performs the following storage-related operations:

- Stores a rating for a given record (identified by a combination of a record ID and its type)
- Gets all ratings for a given record

Let's review the data schema of a rating:

- **User ID**: String

- **Record ID**: String

- **Record type**: String

- **Rating value**: Integer

At this point, we know which data we want to store in the database and can set up our database. We will use a Docker version of MySQL that you can run by executing the following command:

```
docker run --name movieexample_db -e MYSQL_ROOT_
PASSWORD=password -e MYSQL_DATABASE=movieexample -p 3306:3306
-d mysql:latest
```

In our command, we set the password for the MySQL root user to `password` so that we can use it for testing. We also set the database name to `movieexample` and exposed it on port `3306` so that we can use it to access our MySQL database.

Let's verify that our container started successfully. Run the following command to see the list of running Docker containers:

```
docker ps
```

The output should include a container with a `movieexample_db` name, a `mysql:latest` image, and an Up status.

The next step is to create our data schema. We will define it in a separate folder in our `src` directory, called `schema`. Create this directory and a `schema.sql` file in it, and add the following code to the newly created file:

```
CREATE TABLE IF NOT EXISTS movies (id VARCHAR(255), title
VARCHAR(255), description TEXT, director VARCHAR(255));
CREATE TABLE IF NOT EXISTS ratings (record_id VARCHAR(255),
record_type VARCHAR(255), user_id VARCHAR(255), value INT);
```

In our schema file, we define two tables, called `movies` and `ratings`. The tables that we just defined consist of VARCHAR(255) and TEXT columns. VARCHAR is a MySQL type for storing string data, and 255 is the maximal size of a column value. TEXT is another MySQL type that is often used for storing long text records, so we used it for storing movie descriptions that may contain long texts.

Now, let's connect to our newly provisioned database and initialize our data schema. Run the following command inside the `src` directory of our project:

```
docker exec -i movieexample_db mysql movieexample -h localhost
-P 3306 --protocol=tcp -uroot -ppassword < schema/schema.sql
```

If everything worked correctly, our database should be ready to use. You can check if the tables were created successfully by running the following command:

```
docker exec -i movieexample_db mysql movieexample -h localhost
-P 3306 --protocol=tcp -uroot -ppassword -e "SHOW tables"
```

The output of the preceding command should include our two tables:

```
Tables_in_movieexample
movies
ratings
```

We are ready to implement the logic to write and read from it. Create a metadata/internal/ repository/mysql directory and add a file called mysql.go to it with the following contents:

```go
package mysql

import (
    "context"
    "database/sql"
    "metadata/pkg/model"

    _ "github.com/go-sql-driver/mysql"
    "movieexample.com/metadata/internal/repository"
    "movieexample.com/metadata/pkg/model"
)

// Repository defines a MySQL-based movie matadata repository.
type Repository struct {
    db *sql.DB
}

// New creates a new MySQL-based repository.
func New() (*Repository, error) {
    db, err := sql.Open("mysql", "root:password@/movieexample")
    if err != nil {
        return nil, err
```

```
    }
    return &Repository{db}, nil
}
```

In our code, we defined a MySQL-based repository that we will use to store and retrieve the movie metadata. Note that we added the following line to our imports:

```
_   "github.com/go-sql-driver/mysql"
```

The line that we added initializes a Go MySQL **driver**, which is required to access our MySQL database. We also used root:password@/movieexample inside our New function – the value is called a **connection string** and it includes the name of the user, its password, and the name of the database to be connected. The connection string may also include the name of the host, MySQL port, and other values, but we don't need to set them as we are using the default values to access the local version of MySQL.

> **Important Note**
>
> Please note that storing database credentials in code is a bad practice, and it is recommended to store such data (often called *secrets*) separately: for example, as separate configuration files. In *Chapter 8*, we will review how to create and use configuration files with Go microservices.

Now, add the following code to the file that we just created:

```go
// Get retrieves movie metadata for by movie id.
func (r *Repository) Get(ctx context.Context, id string)
(*model.Metadata, error) {
    var title, description, director string
    row := r.db.QueryRowContext(ctx, "SELECT title,
description, director FROM movies WHERE id = ?", id)
    if err := row.Scan(&title, &description, &director); err !=
nil {
        if err == sql.ErrNoRows {
            return nil, repository.ErrNotFound
        }
        return nil, err
    }
    return &model.Metadata{
        ID:          id,
        Title:       title,
        Description: description,
```

```
            Director:     director,
        }, nil
    }


    // Put adds movie metadata for a given movie id.
    func (r *Repository) Put(ctx context.Context, id string,
    metadata *model.Metadata) error {
        _, err := r.db.ExecContext(ctx, "INSERT INTO movies (id,
    title, description, director) VALUES (?, ?, ?, ?)",
            id, metadata.Title, metadata.Description, metadata.
    Director)
        return err
    }
```

In our code, we implemented the Get and Put functions so that we can store and retrieve the movie metadata from MySQL. Inside our Get function, we use the QueryRowContext function of our database instance to read a single row from our table. In the case of a query error, we check if it is equal to sql.ErrNoRows; if so, we return ErrNotFound.

Now, let's implement our MySQL rating repository. Create a rating/internal/repository/ mysql directory and add a mysql.go file to it with the following contents:

```
package mysql

import (
    "context"
    "database/sql"
    "rating/pkg/model"

    _ "github.com/go-sql-driver/mysql"
    "movieexample.com/rating/internal/repository"
    "movieexample.com/rating/pkg/model"
)

// Repository defines a MySQL-based rating repository.
type Repository struct {
    db *sql.DB
}
```

```go
// New creates a new MySQL-based rating repository.
func New() (*Repository, error) {
    db, err := sql.Open("mysql", "root:password@/movieexample")
    if err != nil {
        return nil, err
    }
    return &Repository{db}, nil
}
```

So far, our rating repository code is similar to the metadata repository. Inside the same file, let's implement two functions to read and write rating data:

```go
// Get retrieves all ratings for a given record.
func (r *Repository) Get(ctx context.Context, recordID model.
RecordID, recordType model.RecordType) ([]model.Rating, error)
{
    rows, err := r.db.QueryContext(ctx, "SELECT user_id,
value FROM ratings WHERE record_id = ? AND record_type = ?",
recordID, recordType)
    if err != nil {
        return nil, err
    }
    defer rows.Close()
    var res []model.Rating
    for rows.Next() {
        var userID string
        var value int32
        if err := rows.Scan(&userID, &value); err != nil {
            return nil, err
        }
        res = append(res, model.Rating{
            UserID: model.UserID(userID),
            Value:  model.RatingValue(value),
        })
    }
    if len(res) == 0 {
        return nil, repository.ErrNotFound
    }
}
```

```
        return res, nil
}


// Put adds a rating for a given record.
func (r *Repository) Put(ctx context.Context, recordID model.
RecordID, recordType model.RecordType, rating *model.Rating)
error {
    _, err := r.db.ExecContext(ctx, "INSERT INTO ratings
(record_id, record_type, user_id, value) VALUES (?, ?, ?, ?)",
        recordID, recordType, rating.UserID, rating.Value)
    return err
}
```

In our Get handler, we use the Query function to read rating rows from our table. We scan each row by calling the rows.Scan function, converting MySQL data into the necessary structures.

Our repository code is ready, so we can import the newly used package, github.com/go-sql-driver/mysql, by running the following command:

```
go mod tidy
```

Let's verify that our logic is correct by manually testing the rating repository:

1. Inside the rating/cmd/main.go file, change the movieexample.com/rating/internal/repository/memory import to movieexample.com/rating/internal/repository/mysql.

2. Inside the same file, find the following block:

```
repo := memory.New()
```

3. Change it to the following:

```
repo := mysql.New()
if err != nil {
    panic(err)
}
```

4. Navigate to the cmd directory of the rating service and run the following command:

```
go run *.go
```

5. Make a manual request to write a rating:

```
grpcurl -plaintext -d '{"record_id":"1", "record_type":
"movie"}' localhost:8082 RatingService/GetAggregatedRating
```

You should see the following message:

```
ERROR:
  Code: NotFound
  Message: ratings not found for a record
```

6. Now, let's write a rating to test that our database works correctly. Execute the following command:

```
grpcurl -plaintext -d '{"record_id":"1", "record_type":
"movie", "user_id": "alex", "rating_value": 5}' localhost:8082
RatingService/PutRating
```

7. Now, let's fetch an updated rating for the same movie. Execute the same command as in *step 4*:

```
grpcurl -plaintext -d '{"record_id":"1", "record_type":
"movie"}' localhost:8082 RatingService/GetAggregatedRating
```

8. You should get the following response:

```
{
  "ratingValue": 5
}
```

Hooray, we just confirmed that our repository logic works! You can now shut down the rating service, rerun it, and repeat *step 6*. When you do this, you will get the same result, and this will confirm that our data is persistent now and does not get impacted by service restarts.

Summary

In this chapter, we provided a brief overview of database storage solutions for storing microservice data. We illustrated how to write the logic for writing and reading our service data into MySQL, a popular open source relational database that is widely used across the software development industry.

In the next chapter, we are going to illustrate how to build and run our service instances using a popular platform, Kubernetes, that allows us to coordinate various service-related operations, such as code updates, automated service instance count increases, and many more.

Further reading

To learn more about the topics that were covered in this chapter, take a look at the following resources:

- Types of databases: `https://www.prisma.io/dataguide/intro/comparing-database-types`

- DB engines ranking: `https://db-engines.com/en/ranking`

8

Deployment with Kubernetes

As you have reached this chapter, you already know how to bootstrap microservices, set up the logic for accessing the database, implement service APIs, use serialization, and enable asynchronous communication between your microservices. Now, we are ready to cover a topic that is very important in practice—microservice deployment.

Deployment is a technique of uploading and running your code to one or multiple servers that are often located remotely. Prior to this chapter, we assumed that all services are run locally. We implemented services using static hardcoded local addresses, such as `localhost` for **Kafka**. At some point, you will need to run your services remotely—for example, on a remote server or in a cloud, such as **Amazon Web Services** (**AWS**) or Microsoft Azure.

This chapter will help you to learn how to build and set up your applications for deployments to such remote infrastructure. Additionally, we are going to illustrate how to use one of the most popular deployment and orchestration systems, Kubernetes. You will learn about the benefits it provides, as well as how to set it up for the microservices that we created in the previous chapters.

In this chapter, we will cover the following topics:

- Preparing application code for deployments
- Deploying via Kubernetes
- Deployment best practices

Now, let's proceed to the first part of the chapter, which is going to help you to better understand the core ideas behind the deployment process, and prepare your microservices for deployments.

Technical requirements

To complete this chapter, you need Go `1.11+` or above, similar to the previous chapters. Additionally, you will need Docker, which you can download at `https://www.docker.com`. You will need to register on the Docker website in order to test service deployments in this chapter.

In addition to Docker, to complete this chapter, you will need Kubernetes, which you can download at `https://kubernetes.io` (you will need the `kubectl` and `minikube` tools from it).

You can find the GitHub code for this chapter here:

`https://github.com/PacktPublishing/microservices-with-go/tree/main/Chapter08`

Preparing application code for deployments

In this section, we are going to provide a high-level overview of a service deployment process and describe the actions required to prepare your microservices for deployments. You will learn how to configure Go microservices for running in different environments, how to build them for different operating systems, and some other tips for preparing your microservices for remote execution.

Let's proceed to the basics of the deployment process.

Deployment basics

As we mentioned in the introduction to this chapter, deployments allow you to run and update your applications on one or multiple servers. Such servers are usually located remotely (clouds or dedicated web hosting) and are running all the time to allow your applications to serve the request or process data 24/7.

The deployment process for each environment usually consists of multiple steps. The steps include the following:

1. **Build**: Build a service by compiling it (for compiled languages, such as Go) and including additional required files.

2. **Rollout**: Copy the newly created build to servers of the target environment and replace the existing running code, if any, with the newly built one.

The rollout process is usually sequential: instead of replacing the build on all hosts parallelly, it performs one replacement at a time. For example, if you have ten service instances, the rollout process would first update one instance, then verify that the instance is healthy and move to the second one, and continue until it updates the last service instance. This is done to increase service reliability because if a new version consists of a bug or entirely fails to start on some server, the rollout would not affect all servers at once.

In order to enable the testing of microservices, servers can be classified into multiple categories, called environments:

*   **Local/development**: Servers that are used for running and testing code while working on the code. This environment should never handle any requests from users, and it often consists just of a developer's computer. It can be also configured to use simplified versions of a database and other components, such as single-server and in-memory implementations.

- **Production**: Servers that are intended to handle user requests.

- **Staging**: A mirror of a production environment, but is used for testing. Staging differs from the local/production environment due to configuration and separate data storages, which help to avoid any interference with production data during testing.

Production deployments can be done in **canary** mode—a deployment mode that performs the changes only on a small fraction (such as 1%) of production hosts. Canary deployments are useful for the final testing of new code before updating all production instances of a service.

Let's now see how developers can configure their microservices for deployments to multiple environments.

Application configuration

In the previous section, we described the differences between various environments, such as local/ development and production. Each environment is usually configured differently—if your services have access to databases, each environment will generally have a separate database with different credentials. To enable your services to run in such environments, you would need to have multiple configurations of your services, one per environment.

There are two ways of configuring your services:

- **In-place/hardcode**: All required settings are stored in the service code (Go code, in our case).

- **Separate code and configuration**: Configuration is stored in separate files so that it can be modified independently.

Separating service code and configuration often results in better readability, which makes configuration changes easier. Each environment can have a separate configuration file or a set of files, allowing you to read, review, and update environment-specific configurations easily. Additionally, various data formats, such as YAML, can help to keep configuration files compact. Here's a YAML configuration example:

```
mysql:
  database: ratings
kafka:
  topic: ratings
```

In this book, we are going to use an approach that separates application code and configuration files and stores the configuration in YAML format. This approach is common to many Go applications and can be seen in many popular open source Go projects.

> **Important note**
> Note that invalid configuration changes are among the top causes of service outages in most production systems. I suggest you explore various ways of automatically validating configuration files as a part of the code commit flow. An example of Git-based YAML configuration validation is provided in the following article: `https://ruleoftech.com/2017/git-pre-commit-and-pre-receive-hooks-validating-yaml`.

Let's review our microservice code and see which settings can be extracted from the application configuration:

1. Our `metadata service` does not have any settings other than its gRPC handler address, `localhost:8081`, which you can find in its `main.go` file:

    ```
    lis, err := net.Listen("tcp", fmt.Sprintf("localhost:%v",
    port))
    ```

2. We can extract this setting to the service configuration. A YAML configuration file with this setting would look like this:

    ```
    api:
      port: 8081
    ```

3. Let's make the changes for reading the configuration from a file. Inside the `metadata/cmd` directory, create a `config.go` file and add the following code to it:

    ```
    package main

    type serviceConfig struct {
      APIConfig apiConfig `yaml:"api"`
    }

    type apiConfig struct {
      Port string `yaml:"port"`
    }
    ```

4. In addition to this, create a `configs` directory inside the `metadata` service directory and add a `base.yaml` file to it with the following contents:

    ```
    api:
      port: 8081
    ```

5. The file we just created contains the YAML configuration for our service. Now, let's add code to our `main.go` file to read the configuration. Replace the first line of the `main` function that prints a log message with this:

```
log.Println("Starting the movie metadata service")
f, err := os.Open("base.yaml")
if err != nil {
    panic(err)
}
defer f.Close()
var cfg serviceConfig
if err := yaml.NewDecoder(f).Decode(&cfg); err != nil {
    panic(err)
}
```

Additionally, replace the line with the `net.Listen` call with this:

```
lis, err := net.Listen("tcp", fmt.Sprintf("localhost:%d",
cfg.      APIConfig.Port))
```

6. The code we have just added is using a `gopkg.in/yaml.v3` package to read a YAML file. Import it into our module by running the following command:

```
go mod tidy
```

Make the same changes that we just made for the other two services we created earlier. Use port number `8082` for the `rating` service and `8083` for the `movie` service in your YAML files.

The changes we just made helped us introduce the application configuration that is separate from the service logic. This can help us when we want to introduce additional configurable options—to make any configuration changes, we would just need to update the YAML files without touching our service Go code.

Now that we have finished configuring our microservices for deployment, we are ready to move to the next section, which is going to cover the deployment process of our microservices.

Deploying via Kubernetes

In this section, we are going to illustrate how to set up deployments for our microservices using a popular open source deployment and orchestration platform, Kubernetes. You will learn the basics of Kubernetes, how to set up our microservices for using it, and how to test our microservice deployments in Kubernetes.

Introduction to Kubernetes

Kubernetes is an open source deployment and orchestration platform that was initially created at Google and later maintained by a large developer community backed by the Linux Foundation. Kubernetes provides a powerful, scalable, and flexible solution for running and deploying applications of any size, from small single-instance applications to ones having tens of thousands of instances. Kubernetes helps to orchestrate multiple operations, such as deployments, rollbacks, up- and down-scaling of applications (changing the application instance count upward and downward), and many more.

In Kubernetes, each application consists of one or multiple **pods**—the smallest deployable units. Each pod contains one or multiple **containers**—lightweight software blocks containing the application code. The deployment of a single container to multiple pods is illustrated in the following diagram:

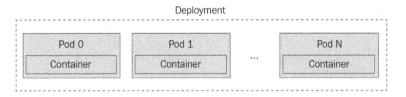

Figure 8.1 – Kubernetes deployment model

Kubernetes pods can be run on one or multiple hosts, called **nodes**. A group of nodes is called a **cluster**, and the relationship between the cluster, nodes, and its pods is illustrated in the following diagram:

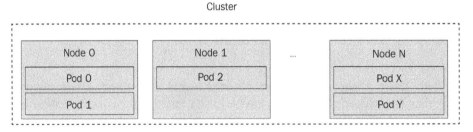

Figure 8.2 – Kubernetes cluster model

For deploying a service in Kubernetes, developers generally need to perform the following steps:

1. **Prepare a container image**: A **container image** contains either the application code or its compiled binary (both options can be used, as long as the container image contains the instructions and any tools to run the code), as well as any additional files required for running it. A container image is essentially a program ready for deployment.

2. **Create a deployment configuration**: A Kubernetes deployment configuration tells it how to run the application. It includes settings such as the number of replicas (number of pods to run), names of containers, and many more.

3. **Run a deployment command**: Kubernetes will apply the provided configuration by running the desired number of pods with the target application(s).

One of the benefits of Kubernetes is abstracting away all the low-level details of deployments, such as selecting target servers to deploy (if you have many, you need to balance their load otherwise), copying and extracting your files, and running health checks. In addition to this, there are some other useful benefits:

- **Service discovery**: Kubernetes offers a built-in service discovery API for use in applications.

- **Rollbacks**: In case there are any issues with the deployment, Kubernetes allows you to roll back the changes to the previous state.

- **Automated restarts**: If any pod experiences any issue, such as an application crash, Kubernetes will perform a restart of that pod.

Now, let's describe how we can set up deployments of our microservices using Kubernetes.

Setting up our microservices for Kubernetes deployments

All the necessary steps for setting up deployments in Kubernetes for our three microservices are set out here:

1. The first step is to create a container image for each service. Kubernetes supports multiple types of containers, and Docker is currently the most popular container type. We already used Docker in *Chapter 3* and will illustrate now how to use it for creating containers for our services.

   Inside the `metadata` service directory, create a file called `Dockerfile` and add the following code to it:

   ```
   FROM alpine:latest


   COPY main .
   COPY configs/. . EXPOSE 8081
   CMD ["/main"]
   ```

   In the file that we just added, we specified that to prepare the image for our container for the `metadata` service, Docker should use the `alpine:latest` base image. **Alpine** is a lightweight Linux distribution that has a size of just a few megabytes and is optimal for our services. Then, we added a command to copy the executable file called `main` to a container, copy the `configs` directory of the service, and expose an `8081` port so that we can accept incoming requests on it.

2. As the next step, add a file with the same contents inside the `rating` and the `movie` service directories. Make sure you use the right ports in the files (`8082` and `8083`, correspondingly).

    Once you have created the Docker configuration files, run the `build` command inside each service directory:

```
GOOS=linux go build -o main cmd/*.go
```

The results of the previous command should be the executable file called `main`, stored in each service directory. Note that we used a `GOOS=linux` variable—this tells the `go` tool to build our code for the Linux operating system.

3. The next step is to build service images. Run this command from the `metadata` service directory:

```
docker build -t metadata .
```

Similarly, run this command from the `rating` service directory:

```
docker build -t rating .
```

Finally, run this command from the `movie` service directory:

```
docker build -t movie .
```

If each command is executed successfully, we are ready to run out containers using the following commands:

```
docker run -p 8081:8081 -it metadata
docker run -p 8082:8082 -it rating
docker run -p 8083:8083 -it movie
```

The result of each execution should be a successful execution of each service.

4. The next step is to create Docker Hub repositories in your account so that you can publish your service images to them. Log in to `https://hub.docker.com`, go to the **Repositories** section, and create three repositories, called `metadata`, `rating`, and `movie`.

    Execute the following commands to publish the images:

```
docker tag metadata <Your Docker username>/metadata:1.0.0
docker push <Your Docker username>/metadata:1.0.0
docker tag metadata <Your Docker username>/rating:1.0.0
docker push <Your Docker username>/rating:1.0.0
docker tag metadata <Your Docker username>/movie:1.0.0
docker push <Your Docker username>/movie:1.0.0
```

These commands should upload the images we just created to your Docker Hub repositories so that Kubernetes can download them during the deployment.

At this point, we are ready to create a Kubernetes deployment configuration that is going to tell Kubernetes how to deploy our services.

5. Inside the `metadata` service directory, create a file called `kubernetes-deployment.yml` with the following contents:

```yaml
apiVersion: apps/v1
kind: Deployment
metadata:
  name: metadata
spec:
  replicas: 2
  selector:
    matchLabels:
      app: metadata
  template:
    metadata:
      labels:
        app: metadata
    spec:
      containers:
      - name: metadata
        image: microservices-with-go/metadata:1.0.0
        imagePullPolicy: IfNotPresent
        ports:
          - containerPort: 8081
```

The file that we just created provides instructions to Kubernetes on how to deploy our service. Here are some important settings:

- **Replicas**: The number of pods to run
- **Image**: The name of the container image to deploy
- **Ports**: Container port to expose

Note that the container port is different from the application port (the one that we configured in our `APIConfig` structure). The mapping between these settings is done by Docker as a part of the `docker run` settings.

6. Now, create a file with the same name in the `rating` service directory with the following contents:

```
apiVersion: apps/v1
kind: Deployment
metadata:
  name: rating
spec:
  replicas: 2
  selector:
    matchLabels:
      app: rating
  template:
    metadata:
      labels:
        app: rating
    spec:
      containers:
      - name: rating
        image: <Your Docker username>/rating:1.0.3
        imagePullPolicy: IfNotPresent
        ports:
          - containerPort: 8082
```

Remember to replace the `image` property with the Docker image name that you created in *step 4*.

7. Finally, create a `kubernetes-deployment.yml` file in the `movie` service directory with the following contents:

```
apiVersion: apps/v1
kind: Deployment
metadata:
  name: movie
spec:
  replicas: 2
  selector:
    matchLabels:
      app: movie
  template:
```

```
metadata:
  labels:
    app: movie
spec:
  containers:
  - name: movie
    image: ashuiskov/movie:1.0.0
    imagePullPolicy: IfNotPresent
    ports:
      - containerPort: 8083
```

8. The next step is to start the local Kubernetes cluster using the `minikube` tool, which you should have installed as a part of Kubernetes. Run the following command to start the cluster:

minikube start

9. Then, apply our `metadata` deployment configuration by running the following command from the `metadata` service directory:

kubectl apply -f kubernetes-deployment.yml

10. If the previous command is executed successfully, you should see the new deployment by running this command:

kubectl get deployments

The output of the command should be this:

```
NAME       READY   UP-TO-DATE   AVAILABLE   AGE
metadata   0/2     2            0           6s
```

Also, check the state of the service pods by running the following command:

kubectl get pods

The output should show the `Running` status for our `metadata` service pods, as shown here:

```
NAME                        READY   STATUS    RESTARTS   AGE
metadata-5f87cbbf65-st69m   1/1     Running   0          116s
metadata-5f87cbbf65-t4xsk   1/1     Running   0          116s
```

As you may notice, Kubernetes created two pods for our service, the same number as we specified in the deployment configuration. Each pod has a **unique identifier** (**UID**), which is shown in the left column. You can see that Kubernetes created two pods for our `metadata` service.

You can check the logs of each pod by running the following command:

```
kubectl logs -f <POD_ID>
```

Now, perform the same changes that we did for the `metadata` service for the other two services, and verify that the pods are running.

If you want to make some manual API requests to the services, you need to set up port forwarding by running the following command:

```
kubectl port-forward <POD_ID> 8081:8081
```

This command would work for the `metadata`, `rating`, and `movie` services; however, you would need to replace the `8081` port value with `8082` and `8083`, correspondingly.

If you did everything well, congratulations! We have finished setting up basic Kubernetes deployments of our microservices. Let's summarize what we did in this section:

- First, we created container images for each of our services so that we could deploy them.
- Then, we published our container images to Docker Hub so that Kubernetes could pull the images during the deployment.
- We created a Kubernetes deployment configuration to tell it how to deploy our microservices.
- Finally, we tested our Kubernetes deployments using a combination of `minikube` and `kubectl` commands.

At this point, you should have some understanding of Kubernetes deployments and know how to deploy your microservices using them. This knowledge will help you to run your services on many platforms, including all popular cloud platforms, such as AWS, Azure, and **Google Cloud Platform** (**GCP**).

Deployment best practices

In this section, we are going to describe some best practices related to the deployment process. These practices, listed here, will help you to set up a reliable deployment process for your microservices:

- Automated rollbacks
- Canary deployments
- Continuous deployment (CD)

Automated rollbacks

Automated rollbacks are the mechanism of automatically reverting a deployment in case there was a failure during it. Imagine you are making deployment of a new version of your service and that version has some application bug that is preventing it from starting successfully. In that case, the deployment process will replace your active instances of a service (if the service is already running) with the failing ones, making your services unavailable. Automated rollbacks are a way to detect and revert such bad deployments, helping you to avoid an outage in situations when your services become unavailable due to such issues.

Automated rollbacks are not offered by default in Kubernetes, at the time of writing this book, similar to many popular deployment platforms. However, this should not stop you from using this technique, especially if you aim to achieve high reliability of your services. The high-level idea of implementing automated rollbacks with Kubernetes is as follows:

- Perform continuous health checks of your service (we are going to cover such logic in *Chapter 12* of this book).

- When you detect a health issue with your service, check whether there was a recent deployment of your service. For example, you can do so by running the `kubectl describe deployment` command.

- In case there was a recent deployment and the time of it closely matches the time when the health check issues were detected, you can roll it back by executing this rollback command: `kubectl rollout undo deployment <DEPLOYMENT_NAME>`.

Canary deployments

As we mentioned at the beginning of the chapter, canary is a special type of deployment, where you update only a small fraction (1 to 3%) of instances. The idea of canary deployments is to test a new version of your code on a subset of production instances and validate its correctness before doing a regular production deployment.

We won't cover the details of setting up canary deployments in Kubernetes, but can cover the basic ideas that would help you to do this once you want to enable canary deployments for your microservices, as set out here:

- Create two separate Kubernetes deployment configurations, one for canary and one for production.

- Specify the desired number of replicas in each configuration—if you want to run a service on 50 pods and let canary handle 2% of traffic, set 1 replica for canary and 49 replicas for production.

- You may also add environment-specific suffixes to deployment names. For example, you can call a canary deployment of a rating service, `rating-canary,` and `rating-production` for the production environment.

- When you perform a deployment of your service, deploy it using a canary configuration first.

- Once you verify that the deployment was successful, make a deployment using a production configuration.

Canary deployments are strongly recommended for increasing the reliability of your deployments. Testing new changes on a small fraction of traffic helps to reduce the impact of various application bugs and other types of issues that your services can encounter.

Replace with Continuous Deployment (CD)

Continuous Deployment (CD) is a technique of making frequent recurring deployments. With CD, services get deployed automatically—for example, on each code change. The main benefit of CD is early deployment failure detection—if any change (such as a Git commit of a new service code) is causing a deployment failure, the failure would often get detected much sooner than in the case of manual deployments.

You can automate deployments by programmatically monitoring a change log (such as Git commit history), or by using **Git hooks**—configurable actions that are executed at specific stages of Git changes. With Kubernetes, once you detect a new version of your software, you can trigger a new deployment by using a `kubectl apply` command.

Due to the high cadence of version updates, CD requires some tooling for automated checks of service health. We are going to cover such tooling later in *Chapter 11* and *Chapter 12* of this book.

Summary

In this chapter, we have covered a very important topic—service deployments. You have learned about the basics of the service deployment process, as well as the necessary steps for preparing our microservices for deployment. Then, we introduced Kubernetes, a popular deployment and orchestration platform that is now provided by many companies and cloud providers. We have illustrated how to set up a local Kubernetes cluster and deploy our microservices to it, running multiple instances of each service to illustrate how easy is to run any arbitrary number of instances within the Kubernetes platform.

The knowledge you gained should help you to set up more complex deployment processes, as well as to work with the services that are already deployed via Kubernetes.

This chapter summarizes our material on service deployments. In the next chapter, we are going to describe another important topic: unit and integration.

Further reading

If you'd like to learn more, refer to the following links:

- Kubernetes documentation: `https://kubernetes.io/docs/home/`

- Service deployment best practices: `https://codefresh.io/learn/software-deployment/`

- Setting up Kubernetes services: `https://kubernetes.io/docs/concepts/services-networking/service/`

- Blue-green deployments: `https://www.redhat.com/en/topics/devops/what-is-blue-green-deployment`

Unit and Integration Testing

Testing is an integral part of any development process. It is always important to cover your code with automated tests, ensuring that all important logic is continuously tested on all code changes. Writing good tests often helps ensure that any changes made throughout the development process will keep the code working and reliable.

Testing is especially important in microservice development, but it brings some additional challenges to developers. It's not enough to test each service – it's also important to test the integrations between the services, ensuring every service can work with the others.

In this chapter, we will cover both unit testing and integration testing and illustrate how to add tests to the microservices we created in the previous chapters. We will cover the following topics:

- Go testing overview
- Unit tests
- Integration tests
- Testing best practices

You will learn how to write unit and integration tests in Go, how to use the mocking technique, and how to organize the testing code for your microservices. This knowledge will help you to build more reliable services.

Let's proceed to the overview of Go testing tools and techniques.

Technical requirements

To complete this chapter, you need Go version 1.11+ or above.

You can find the GitHub code for this chapter here: `https://github.com/PacktPublishing/microservices-with-go/tree/main/Chapter09`.

Go testing overview

In this section, we are going to provide a high-level overview of Go's testing capabilities. We will cover the basics of writing tests for Go code, list the useful functions and libraries provided with the Go SDK, and describe various techniques for writing tests that will help you in microservice development.

First, let's cover the basics of writing tests for Go applications.

Go language has built-in support for writing automated tests and provides a package called `testing` for this purpose.

There is a conventional relationship between the Go code and its tests. If you have a file called `example.go`, its tests would reside in the same package in a file called `example_test.go`. Using a `_test` file name suffix allows you to differentiate between the code being tested and the tests for it, making it easier to navigate the source code.

Go test functions follow this conventional name format, with each test function name starting with the `Test` prefix:

```
func TestXxx(t *testing.T)
```

Inside these functions, you can use the `testing.T` structure to report test failures or use any additional helper functions provided by it.

Let's take this test as an example:

```
func TestAdd(t *testing.T) {
  a, b := 1, 2
  if got, want := Add(1, 2), 3; got != want {
    t.Errorf("Add(%v, %v) = %v, want %v", a, b, got, want)
  }
}
```

In the preceding function, we used `testing.T` to report a test failure in case the `Add` function provides an unexpected output.

When it comes to execution, we can run the following command:

```
go test
```

The command executes each test in the target directory and prints the output, containing error messages for any failing tests or any other necessary data.

Developers are free to choose the format of their tests; however, there are some common techniques, such as **table-driven tests**, that often help organize test code elegantly.

Table-driven tests are tests in which inputs are stored in the form of a table or a set of rows. Let's take this example:

```
func TestAdd(t *testing.T) {
    tests := []struct {
        a    int
        b    int
        want int
    }{
        {a: 1, b: 2, want: 3},
        {a: -1, b: -2, want: -3},
        {a: -3, b: 3, want: 0},
        {a: 0, b: 0, want: 0},
    }
    for _, tt := range tests {
        assert.Equal(t, tt.want, Add(tt.a, ,tt.b), fmt.
Sprintf("Add(%v, %v)", tt.a, tt.b))
    }
}
```

In this code, we initialize the `tests` variable with the test cases for our function and then iterate over it. Note that we use the `assert.Equal` function provided by the `github.com/stretchr/testify` library to compare the expected and the actual result of the function being tested. This library provides a set of convenient functions that can simplify your test logic. Without using the `assert` library, the code comparing the test result would look like the following:

```
        if got, want := Add(tt.a, tt.b), tt.want; got != want {
            t.Errorf ("Add(%v, %v) = %v, want %v", tt.a, tt.b,
got, want)
        }
```

Table-driven tests help reduce the repetitiveness of tests by separating test cases and the logic that performs actual checks. In general, these tests are good practice when you need to perform lots of similar checks against the defined goal states, as shown in our example.

The table-driven format also helps us improve the readability of test code, making it easier to see and compare different test cases for the same functions. The format is quite common in Go tests; however, you can always organize your test code in the way that is the best for your use case.

Now, let's review the basic features provided by Go's built-in testing library.

Subtests

One of the interesting features of the Go testing library is the ability to create **subtests** — tests that get executed inside other ones. Among the benefits of subtests is the ability to execute them separately, as well as to execute them in parallel for long-running tests and structure the test output in a more granular way.

Subtests are created by calling the Run function of the testing library:

```
func (t *T) Run(name string, f func(t *T)) bool
```

When using the Run function, you need to pass the name of the test case and the function to execute, and Go will take care of executing each test case separately. Here's an example of a test using the Run function:

```
func TestProcess(t *testing.T) {
  t.Run("test case 1", func(t *testing.T) {
    // Test case 1 logic.
  })
  t.Run("test case 2", func(t *testing.T) {
    // Test case 2 logic.
  })
}
```

In the preceding example, we created two subtests by calling the Run function twice, one time for each subtest.

To achieve more fine-grained control over subtests, you can use the following options:

- Each subtest, either passing or failing, can be shown separately in the output when running the go test command with the -v argument
- You can run an individual test case by using a -run argument of the go test command

There is one other interesting benefit of using the Run function. Let's imagine that you have a function called Process that takes seconds to complete. If you have a table test with lots of test cases and you execute them sequentially, the execution of the entire test may take a lot of time. In this case, you could let the Go test runner execute tests in parallel mode by calling the t.Parallel() function. Let's illustrate this in the following example:

```
func TestProcess(t *testing.T) {
    tests := []struct {
        name    string
        input   string
```

```
            want    string
      }{
          {name: "empty", input: "", want: ""},
          {name: "dog", input: "animal that barks", want: "dog"},
          {name: "cat", input: "animal that meows", want: "cat"},
      }
      for _, tt := range tests {
          input := tt.input
          t.Run(tt.name, func(t *testing.T) {
              t.Parallel()
              assert.Equal(t, tt.want, Process(input), fmt.
Sprintf("Process(%v)", input))
          })
      }
}
```

In our example, we call the t.Run function for each test case, passing the test case name and the function to be executed. Then, we call t.Parallel() to make each test case execute in parallel. This optimization would significantly reduce the execution time in the case that our Process function is very slow.

Skipping

Imagine that you want to execute your Go tests after each change on your computer, but you have some slow tests that take a long time to run. In that case, you would want to find a way to skip running tests under certain conditions. The Go testing library has built-in support for this – the Skip function. Let's take this test function as an example:

```
func TestProcess(t *testing.T) {
  if os.Getenv("RUNTIME_ENV") == "development" {
    t.Skip("Skipping a test in development environment")
  }
  ...
}
```

In the preceding code, we skip the test execution if there is a RUNTIME_ENV runtime environment variable with the development value. Note that we also provide the reason for skipping it inside the t.Skip call so that it is logged on test execution.

The skipping feature can be particularly useful for bypassing the execution of long-running tests, such as the tests performing slow I/O operations or doing lots of data processing. To support this, the Go testing library provides an ability to pass a specific flag, -test.short, to the go test command:

```
go test -test.short
```

With the -test.short flag, you can let the Go test runner know that you want to run tests in **short mode** — a testing mode when only special, short tests are getting executed. You can add the following logic to all long-running tests to exclude them in short mode:

```
func TestLongRunningProcess(t *testing.T) {
  if testing.Short() {
    t.Skip("Skipping a test in short mode")
  }
  ...
}
```

In the preceding example, the test is skipped when the -test.short flag is passed to the test command.

Using the short testing mode is useful when some of your tests are much slower than others and you need to run tests very frequently. Skipping the slow tests and executing them less frequently could significantly increase your development speed and make your development experience much better.

You can get familiar with the other Go testing features by checking out the official documentation for the testing package: https://pkg.go.dev/testing. We are now going to proceed to the next section and focus on the details of implementing unit tests for our microservices.

Unit tests

We have covered many useful features for automated testing of Go applications and are now ready to illustrate how to use them in our microservice code. First, we are going to start with **unit tests** — tests of individual units of code, such as structures and individual functions.

Let's walk through the process of implementing unit tests for our code using the metadata service controller as an example. Currently, our controller file looks like this:

```
package metadata

import (
    "context"
```

```
        "movieexample.com/metadata/pkg/model"
)


type metadataRepository interface {
    Get(ctx context.Context, id string) (*model.Metadata,
error)
}


// Controller defines a metadata service controller.
Type Controller struct {
    repo metadataRepository
}


// New creates a metadata service controller.
Func New(repo metadataRepository) *Controller {
    return &Controller{repo}
}


// Get returns movie metadata by id.
Func (c *Controller) Get(ctx context.Context, id string)
(*model.Metadata, error) {
    return c.repo.Get(ctx, id)
}
```

Let's list what we would like to test in our code:

- A Get call when the repository returns ErrNotFound
- A Get call when the repository returns an error other than ErrNotFound
- A Get call when the repository returns metadata and no error

So far, we have three test cases to implement. All test cases need to perform operations on the metadata repository and we need to simulate three different responses from it. How exactly should we simulate the responses from our metadata repository in the test? Let's explore the powerful technique that allows us to achieve this with our testing code.

Mocking

The technique of simulating responses from a component is called **mocking**. Mocking is often used in tests to simulate various scenarios, such as returning specific results or errors. There are multiple ways of using mocking in Go code. The first one is to implement the *fake* version of components, called **mocks**, manually. Let's illustrate how to implement these mocks using our metadata repository as an example. Our metadata repository interface is defined in the following way:

```go
type metadataRepository interface {
    Get(ctx context.Context, id string) (*model.Metadata,
error)
}
```

The mock implementation of this interface could look like this:

```go
type mockMetadataRepository struct {
    returnRes *model.Metadata
    returnErr error
}

func (m *mockMetadataRepository) setReturnValues(res *model.
Metadata, err error) {
    m.returnRes = res
    m.returnErr = err
}

func (m *mockMetadataRepository) Get(ctx context.Context, id
string) (*model.Metadata, error) {
    return m.returnRes, m.returnErr
}
```

In our example mock of the metadata repository, we allow set values to be returned on the upcoming calls to the Get function by providing the setReturnValues function. The mock could be used to test our controller in the following way:

```go
m := mockMetadataRepository{}
m.setReturnValues(nil, repository.ErrNotFound)
c := New(m)
res, err := c.Get(context.Background(), "some-id")
// Check res, err.
```

Manual implementation of mocks is a relatively simple way to test calls to various components that are outside of the scope of the package being tested. The downside of this approach is that you need to write mock code by yourself and update its code on any interface changes.

The other way of using mocks is to use libraries that generate mocking code. An example of this kind of library is `https://github.com/golang/mock`, which contains a mock generation tool called `mockgen`. You can install it by running the following command:

```
go install github.com/golang/mock/mockgen
```

The `mockgen` tool can then be used in the following way:

```
mockgen -source=foo.go [options]
```

Let's illustrate how to generate mock code for our metadata repository. Run the following command from the `src` directory of our project:

```
mockgen -package=repository -source=metadata/internal/
controller/metadata/controller.go
```

You should get the contents of a mock source file as the output. The contents would be similar to this:

```go
// MockmetadataRepository is a mock of metadataRepository
// interface
type MockmetadataRepository struct {
    ctrl     *gomock.Controller
    recorder *MockmetadataRepositoryMockRecorder
}


// NewMockmetadataRepository creates a new mock instance
func NewMockmetadataRepository(ctrl *gomock.Controller)
*MockmetadataRepository {
    mock := &MockmetadataRepository{ctrl: ctrl}
    mock.recorder = &MockmetadataRepositoryMockRecorder{mock}
    return mock
}


// EXPECT returns an object that allows the caller to indicate
// expected use
func (m *MockmetadataRepository) EXPECT()
*MockmetadataRepositoryMockRecorder {
```

```
        return m.recorder
}


// Get mocks base method.
func (m *MockmetadataRepository) Get(ctx context.Context, id
string) (*model.Metadata, error) {
    ret := m.ctrl.Call(m, "Get", ctx, id)
    ret0, _ := ret[0].(*model.Metadata)
    ret1, _ := ret[1].(error)
    return ret0, ret1
}
```

The generated mock code implements our interface and allows us to set the expected responses to our Get function in the following way:

```
ctrl := gomock.NewController(t)
defer ctrl.Finish()
m := NewMockmetadataRepository(gomock.NewController())
ctx := context.Background()
id := "some-id"
m.EXPECT().Get(ctx, id).Return(nil, repository.ErrNotFound)
```

The mock code generated by the gomock library provides some useful features that we have not implemented in our manually created mock version. One of them is the ability to set the expected number of times that the target function should be called using the Times function:

```
m.EXPECT().Get(ctx, id).Return(nil, repository.ErrNotFound).
Times(1)
```

In the preceding example, we limit the number of times the Get function is called to one. The gomock library verifies these constraints at the end of the test execution and reports whether the function was called a different number of times. This mechanism is pretty useful when you want to make sure the target function has definitely been called in your test.

So far, we have shown how to use mocks in two different ways, and you may ask what the preferred way of using them is. Let's compare the two approaches to find out the answer.

The benefit of implementing mocks manually is the ability to do so without using any external libraries, such as gomock. However, the downsides of this approach would be the following:

- Manual implementation of mocks takes time

- Any changes to the mocked interfaces would require manual updates to the mock code

- Harder to implement extra features that are provided by libraries such as gomock, such as call count verification

Using a library such as gomock for providing mock code would be beneficial for the following reasons:

- Higher code consistency when all mocks are generated in the same way

- No need to write boilerplate code

- An extended mock feature set

In our comparison, automatic mock code generation seems to provide more advantages, so we will follow the gomock-based approach for automatic mock generation. In the next section, we are going to demonstrate how to do this for our services.

Implementing unit tests

We are going to illustrate how to implement controller unit tests using the generated gomock code. First, we will need to find a good place in our repository to put the generated code. We already have a directory called gen that is shared among the services. We can create a sub-directory called mock that we can use for various generated mocks. Run the mock generation command for the metadata repository again:

```
mockgen -package=repository -source=metadata/internal/
controller/metadata/controller.go
```

Copy its output to the file called gen/mock/metadata/repository/repository.go. Now, let's add a test for our metadata service controller. Create a file called controller_test.go in its directory and add to it the following code:

```
package metadata

import (
    "context"
    "errors"
    "testing"

    "github.com/golang/mock/gomock"
    "github.com/stretchr/testify/assert"
    gen "movieexample.com/gen/mock/metadata/repository"
    "movieexample.com/metadata/internal/repository"
```

```
        "movieexample.com/metadata/pkg/model"
)
```

Then, add the following code, containing the test cases in a table format:

```
func TestController(t *testing.T) {
    tests := []struct {
        name       string
        expRepoRes *model.Metadata
        expRepoErr error
        wantRes    *model.Metadata
        wantErr    error
    }{
        {
            name:       "not found",
            expRepoErr: repository.ErrNotFound,
            wantErr:    ErrNotFound,
        },
        {
            name:       "unexpected error",
            expRepoErr: errors.New("unexpected error"),
            wantErr:    errors.New("unexpected error"),
        },
        {
            name:       "success",
            expRepoRes: &model.Metadata{},
            wantRes:    &model.Metadata{},
        },
    }
```

Finally, add the code to execute our tests:

```
    for _, tt := range tests {
        t.Run(tt.name, func(t *testing.T) {
            ctrl := gomock.NewController(t)
            defer ctrl.Finish()
            repoMock := gen.NewMockmetadataRepository(ctrl)
            c := New(repoMock)
```

```
            ctx := context.Background()
            id := "id"
            repoMock.EXPECT().Get(ctx, id).Return(tt.
    expRepoRes, tt.expRepoErr)
            res, err := c.Get(ctx, id)
            assert.Equal(t, tt.wantRes, res, tt.name)
            assert.Equal(t, tt.wantErr, err, tt.name)
        })
    }
}
```

The code that we just added implements three different test cases for our Get function using the generated repository mock. We let the mock return the specific values by calling the EXPECT function and passing the desired values. We organized our test in a table-driven way, which we described earlier in the chapter.

To run the tests, use the regular command:

```
go test
```

If you did everything correctly, the output of the test should include ok. Congratulations, we have just implemented the unit tests and demonstrated how to use mocks! We will let you implement the remaining tests for the microservices yourself — it's going to be a fair amount of work, but this is always a great investment for ensuring the code remains tested and reliable.

In the next section, we are going to work on another type of test – integration tests. Knowing why and how to write integration tests in addition to regular unit tests for your microservices will help you to write more stable code and make sure all services work well in integration with each other.

Integration tests

Integration tests are automated tests that verify the correctness of integrations between the individual units of your services and the services themselves. In this section, you are going to learn how to write integration tests and how to structure the logic inside them, as well as get some useful tips that will help you write your own integration tests in the future.

Unlike unit tests that test the individual pieces of code, such as functions and structures, integration tests help ensure that the combinations of individual pieces still work well together.

Let's provide an example of an integration test, taking our rating service as an example. The integration test for our service would instantiate both the service instance and the client for it and ensure that client requests would produce the expected results. As you remember, our rating service provides two API endpoints:

- `PutRating`: Writes a rating to the database
- `GetAggregatedRating`: Retrieves the ratings for a provided record (such as a movie) and returns the aggregated value

Our integration test for the rating service could have the following sequence of calls:

- Writes some data using the `PutRating` endpoint
- Verifies the data using the `GetAggregatedRating` endpoint
- Writes new data using the `PutRating` endpoint
- Calls the `GetAggregatedRating` endpoint and checks that the aggregated value reflects the latest rating update

In microservice development, integration tests usually test individual services or combinations of them – developers can write tests that target an arbitrary number of services.

Unlike unit tests—which generally reside together with the code being tested and can access some internal functions, structures, constants, and variables—integration tests often treat the components being tested as **black boxes**. Black boxes are logical blocks for which the implementation details remain unknown and can only be accessed through publicly exposed APIs or user interfaces. This way of testing is called **black box testing** – testing of a system using a public interface, such as an API, instead of calling individual internal functions or accessing internal components of the system.

Microservice integration tests are often performed by instantiating service instances and performing requests either by calling service APIs or via asynchronous events in case the system handles requests in an asynchronous fashion. The structure of an integration test usually follows a similar pattern:

- **Set up the test**: Instantiate the components being tested and any clients that can access their interfaces
- **Perform test operations and verify the correctness of results**: Run an arbitrary number of operations and compare the outputs from the system being tested, such as a microservice, to the expected values
- **Tear down the test**: Gracefully terminate the test by tearing down the components instantiated in the setup, closing any clients if needed

To illustrate how to write an integration test, let's take three microservices from the previous chapters – metadata, movie, and rating services. To set up our test, we would need to instantiate six components – a server and a client for each microservice. To make it easier to run the test, we can instantiate servers using in-memory implementations of service registries and repositories.

Before you write the test, it's often helpful to write down the set of operations to be tested and determine the expected outputs for each step. Let's write down the plan for our integration test:

1. Write metadata for an example movie using the metadata service API (the `PutMetadata` endpoint) and check that the operation does not return any errors.

2. Retrieve the metadata for the same movie using the metadata service API (the `GetMetadata` endpoint) and check it matches the record that we submitted earlier.

3. Get the movie details (which should only consist of metadata) for our example movie using the movie service API (the `GetMovieDetails` endpoint) and make sure the result matches the data that we submitted earlier.

4. Write the first rating for our example movie using the rating service API (the `PutRating` endpoint) and check the operation does not return any errors.

5. Retrieve the initial aggregated rating for our movie using the rating service API (the `GetAggregatedRating` endpoint) and check that the value matches the one that we just submitted in the previous step.

6. Write the second rating for our example movie using the rating service API and check that the operation does not return any errors.

7. Retrieve the new aggregated rating for our movie using the rating service API and check that the value reflects the last rating.

8. Get the movie details for our example movie and check that the result includes the updated rating.

Having this kind of plan makes it easier to write the code for the integration test and brings us to the last step — actually implementing it:

1. Create a `test/integration` directory and add the file called `main.go` with the following code:

```
package main

import (
    "context"
    "log"
    "net"

    "github.com/google/go-cmp/cmp"
    "github.com/google/go-cmp/cmp/cmpopts"
    "google.golang.org/grpc"
    "movieexample.com/gen"
```

```
        metadatatest "movieexample.com/metadata/pkg/testutil"
        movietest "movieexample.com/movie/pkg/testutil"
        "movieexample.com/pkg/discovery"
        "movieexample.com/pkg/discovery/memory"
        ratingtest "movieexample.com/rating/pkg/testutil"
        "google.golang.org/grpc/credentials/insecure"
)
```

2. Let's add some constants with service names and addresses that we can use later in the test to the file:

```
const (
        metadataServiceName = "metadata"
        ratingServiceName   = "rating"
        movieServiceName    = "movie"

        metadataServiceAddr = "localhost:8081"
        ratingServiceAddr   = "localhost:8082"
        movieServiceAddr    = "localhost:8083"
)
```

3. The next step is to implement the setup code to instantiate our service servers:

```
func main() {
        log.Println("Starting the integration test")

        ctx := context.Background()
        registry := memory.NewRegistry()

        log.Println("Setting up service handlers and
clients")

        metadataSrv := startMetadataService(ctx, registry)
        defer metadataSrv.GracefulStop()
        ratingSrv := startRatingService(ctx, registry)
        defer ratingSrv.GracefulStop()
        movieSrv := startMovieService(ctx, registry)
        defer movieSrv.GracefulStop()
```

Note that `defer` calls to the `GracefulStop` function of each server — this code is a part of the tear-down logic of our test for terminating all servers gracefully.

4. Now, let's set up the test clients for our services:

```
    opts := grpc.WithTransportCredentials(insecure.
NewCredentials())
    metadataConn, err := grpc.Dial(metadataServiceAddr,
opts)
    if err != nil {
        panic(err)
    }
    defer metadataConn.Close()
    metadataClient := gen.
NewMetadataServiceClient(metadataConn)



    ratingConn, err := grpc.Dial(ratingServiceAddr, opts)
    if err != nil {
        panic(err)
    }
    defer ratingConn.Close()
    ratingClient := gen.NewRatingServiceClient(ratingConn)



    movieConn, err := grpc.Dial(movieServiceAddr, opts)
    if err != nil {
        panic(err)
    }
    defer movieConn.Close()
    movieClient := gen.NewMovieServiceClient(movieConn)
```

Now, we are ready to implement the sequence of our test commands. The first step is to test, write, and read the operations of the metadata service:

```
log.Println("Saving test metadata via metadata service")

m := &gen.Metadata{
    Id:            "the-movie",
```

```
        Title:      "The Movie",
        Description: "The Movie, the one and only",
        Director:    "Mr. D",
    }


    if _, err := metadataClient.PutMetadata(ctx, &gen.
PutMetadataRequest{Metadata: m}); err != nil {
        log.Fatalf("put metadata: %v", err)
    }


    log.Println("Retrieving test metadata via metadata
service")


    getMetadataResp, err := metadataClient.GetMetadata(ctx,
&gen.GetMetadataRequest{MovieId: m.Id})
    if err != nil {
        log.Fatalf("get metadata: %v", err)
    }
    if diff := cmp.Diff(getMetadataResp.Metadata, m, cmpopts.
IgnoreUnexported(gen.Metadata{})); diff != "" {
        log.Fatalf("get metadata after put mismatch: %v", diff)
    }
```

You may notice that we used the `cmpopts.IgnoreUnexported(gen.Metadata{})` option inside the call to the `cmp.Diff` function — this tells the `cmp` library to ignore the unexported fields in the `gen.Metadata` structure. We have added this option because the `gen.Metadata` structure, generated by the Protocol Buffers code generator, includes some private fields that we want to ignore in the comparison.

The next test in our sequence would be to retrieve the movie details and check that the metadata matches the record that we submitted earlier:

```
    log.Println("Getting movie details via movie service")


    wantMovieDetails := &gen.MovieDetails{
        Metadata: m,
    }


    getMovieDetailsResp, err := movieClient.
```

```
GetMovieDetails(ctx, &gen.GetMovieDetailsRequest{MovieId:
m.Id})
    if err != nil {
        log.Fatalf("get movie details: %v", err)
    }
    if diff := cmp.Diff(getMovieDetailsResp.MovieDetails,
wantMovieDetails, cmpopts.IgnoreUnexported(gen.MovieDetails{},
gen.Metadata{})); diff != "" {
        log.Fatalf("get movie details after put mismatch: %v",
err)
    }
```

Now, we are ready to test the rating service.

Let's implement two tests – one for writing a rating and one for retrieving the initial aggregated value, which should match the first rating:

```
    log.Println("Saving first rating via rating service")


    const userID = "user0"
    const recordTypeMovie = "movie"
    firstRating := int32(5)
    if _, err = ratingClient.PutRating(ctx, &gen.
PutRatingRequest{
        UserId:      userID,
        RecordId:    m.Id,
        RecordType:  recordTypeMovie,
        RatingValue: firstRating,
    }); err != nil {
        log.Fatalf("put rating: %v", err)
    }


    log.Println("Retrieving initial aggregated rating via
rating service")


    getAggregatedRatingResp, err := ratingClient.
GetAggregatedRating(ctx, &gen.GetAggregatedRatingRequest{
        RecordId:    m.Id,
        RecordType: recordTypeMovie,
```

```
    })
    if err != nil {
        log.Fatalf("get aggreggated rating: %v", err)
    }


    if got, want := getAggregatedRatingResp.RatingValue,
float64(5); got != want {
        log.Fatalf("rating mismatch: got %v want %v", got,
want)
    }
```

The next part of the test would be to submit the second rating and check that the aggregated value was changed:

```
    log.Println("Saving second rating via rating service")


    secondRating := int32(1)
    if _, err = ratingClient.PutRating(ctx, &gen.
PutRatingRequest{
        UserId:      userID,
        RecordId:    m.Id,
        RecordType:  recordTypeMovie,
        RatingValue: secondRating,
    }); err != nil {
        log.Fatalf("put rating: %v", err)
    }


    log.Println("Saving new aggregated rating via rating
service")


    getAggregatedRatingResp, err = ratingClient.
GetAggregatedRating(ctx, &gen.GetAggregatedRatingRequest{
        RecordId:    m.Id,
        RecordType:  recordTypeMovie,
    })
    if err != nil {
        log.Fatalf("get aggreggated rating: %v", err)
    }
```

```
        wantRating := float64((firstRating + secondRating) / 2)
        if got, want := getAggregatedRatingResp.RatingValue,
wantRating; got != want {
            log.Fatalf("rating mismatch: got %v want %v", got,
want)
        }
```

We are almost done with our main function – let's implement the last check:

```
        log.Println("Getting updated movie details via movie
service")

        getMovieDetailsResp, err = movieClient.GetMovieDetails(ctx,
&gen.GetMovieDetailsRequest{MovieId: m.Id})
        if err != nil {
            log.Fatalf("get movie details: %v", err)
        }
        wantMovieDetails.Rating = wantRating
        if diff := cmp.Diff(getMovieDetailsResp.MovieDetails,
wantMovieDetails, cmpopts.IgnoreUnexported(gen.MovieDetails{},
gen.Metadata{})); diff != "" {
            log.Fatalf("get movie details after update mismatch:
%v", err)
        }

        log.Println("Integration test execution successful")
}
```

Our integration test is almost ready. Let's add the functions for initializing the servers for our services below the main function. First, add the function for creating the server for a metadata service:

```
func startMetadataService(ctx context.Context, registry
discovery.Registry) *grpc.Server {
    log.Println("Starting metadata service on " +
metadataServiceAddr)
    h := metadatatest.NewTestMetadataGRPCServer()
    l, err := net.Listen("tcp", metadataServiceAddr)
    if err != nil {
        log.Fatalf("failed to listen: %v", err)
```

```
    }
    srv := grpc.NewServer()
    gen.RegisterMetadataServiceServer(srv, h)
    go func() {
        if err := srv.Serve(l); err != nil {
            panic(err)
        }
    }()
    id := discovery.GenerateInstanceID(metadataServiceName)
    if err := registry.Register(ctx, id, metadataServiceName,
metadataServiceAddr); err != nil {
        panic(err)
    }
    return srv
}
```

You may notice that we call the `srv.Serve` function inside a goroutine — this way, it doesn't block the execution and allows us to immediately return from the function.

Let's add a similar implementation for the rating service server to the same file:

```
func startRatingService(ctx context.Context, registry
discovery.Registry) *grpc.Server {
    log.Println("Starting rating service on " +
ratingServiceAddr)
    h := ratingtest.NewTestRatingGRPCServer()
    l, err := net.Listen("tcp", ratingServiceAddr)
    if err != nil {
        log.Fatalf("failed to listen: %v", err)
    }
    srv := grpc.NewServer()
    gen.RegisterRatingServiceServer(srv, h)
    go func() {
        if err := srv.Serve(l); err != nil {
            panic(err)
        }
    }()
    id := discovery.GenerateInstanceID(ratingServiceName)
```

```
    if err := registry.Register(ctx, id, ratingServiceName,
ratingServiceAddr); err != nil {
        panic(err)
    }
    return srv
}
```

Finally, let's add a function for initializing the movie server:

```
func startMovieService(ctx context.Context, registry discovery.
Registry) *grpc.Server {
    log.Println("Starting movie service on " +
movieServiceAddr)
    h := movietest.NewTestMovieGRPCServer(registry)
    l, err := net.Listen("tcp", movieServiceAddr)
    if err != nil {
        log.Fatalf("failed to listen: %v", err)
    }
    srv := grpc.NewServer()
    gen.RegisterMovieServiceServer(srv, h)
    go func() {
        if err := srv.Serve(l); err != nil {
            panic(err)
        }
    }()
    id := discovery.GenerateInstanceID(movieServiceName)
    if err := registry.Register(ctx, id, movieServiceName,
movieServiceAddr); err != nil {
        panic(err)
    }
    return srv
}
```

Our integration test is ready! You can run it by executing the following command:

```
go run test/integration/*.go
```

If everything is correct, you should see the following output:

```
2022/07/16 16:20:46 Starting the integration test
2022/07/16 16:20:46 Setting up service handlers and clients
2022/07/16 16:20:46 Starting metadata service on localhost:8081
2022/07/16 16:20:46 Starting rating service on localhost:8082
2022/07/16 16:20:46 Starting movie service on localhost:8083
2022/07/16 16:20:46 Saving test metadata via metadata service
2022/07/16 16:20:46 Retrieving test metadata via metadata
service
2022/07/16 16:20:46 Getting movie details via movie service
2022/07/16 16:20:46 Saving first rating via rating service
2022/07/16 16:20:46 Retrieving initial aggregated rating via
rating service
2022/07/16 16:20:46 Saving second rating via rating service
2022/07/16 16:20:46 Saving new aggregated rating via rating
service
2022/07/16 16:20:46 Getting updated movie details via movie
service
2022/07/16 16:20:46 Integration test execution successful
```

As you may notice, the structure of our integration test precisely matches the sequence of test operations that we defined earlier. We implemented our integration test as an executable command and added enough log messages to help you with debugging – if any step fails, it is therefore easier to understand at which step the failure occurred and which operations preceded that step.

It is important to note that we used the in-memory versions of the metadata and rating repositories in our integration test. An alternative approach would be to set up an integration test that stores the data in some persistent databases, such as MySQL. However, there are some challenges with using existing persistent databases in integration tests:

- Integration test data should not interfere with user data. Otherwise, it may cause unexpected effects on existing service users.
- Ideally, test data should be cleaned up after test execution so that the database does not get filled with unnecessary, temporary data.

In order to avoid interference with the existing user data, I would suggest running integration tests on non-production environments, such as staging. Additionally, I would suggest always generating random identifiers for your test records to make sure that individual test executions don't affect each other. For example, you can use the github.com/google/uuid library to generate new identifiers using the uuid.New() function. Lastly, I would recommend always including cleanup code at the end of each integration test that uses persistent data storage to clean up the created records, whenever this is possible.

Now, the question is when we should write integration tests. It is always up to you; however, I do have some general suggestions:

- **Test critical flows**: Make sure you test the entire flows, such as user signups and logins
- **Test critical endpoints**: Perform the tests of the most critical endpoints that your services provide to your users

Additionally, you may have integration tests that are executed after each code change. Systems such as Jenkins provide these kinds of features and allow you to plug any custom logic that would be executed into each update of your code. We won't cover Jenkins setup in this book, but you can familiarize yourself with its documentation on the official website (`https://www.jenkins.io`).

As we have illustrated how to write both unit and integration tests, let's proceed to the next section of the book, describing some of the best practices of Go testing.

Testing best practices

In this section, we are going to list some additional useful testing tips that are going to help you to improve the quality of your tests.

Using helpful messages

One of the most important aspects of writing tests is providing enough information in error logs that it is easy to understand exactly what went wrong and which test case triggered the failure. Consider the following test case code:

```
if got, want := Process(tt.in), tt.want; got != want {
    t.Errorf("Result mismatch")
}
```

The error log does not include both the expected and the actual value received from the function being tested, making it harder to understand what the function returned and how it was different from the expected value.

The better log line would be as follows:

```
t.Errorf("got %v, want %v", got, want)
```

This log line includes the expected and the actual returned value of the function and provides much more context to you when you debug the test.

> **Important note**
>
> Note that in our test logs, first, we log the actual value and then the expected one. This order is recommended by the Go team as the conventional way of logging the values in tests and is followed in all libraries and packages. Follow the same order in your logs for consistency.

An even better error message would be as follows:

```
t.Errorf("YourFunc(%v) = %v, want %v", tt.in, got, want)
```

This error log message includes some additional information – the function being called and the input argument that was passed to it.

To standardize the code for your test cases, you can use the `github.com/stretchr/testify` library. The following example illustrates how to compare the expected and the actual value and log the name of the function being tested, as well as the argument passed to it:

```
assert.Equal(t, want, got, fmt.Sprintf("YourFunc(%v)", tt.in))
```

The assert package of the `github.com/stretchr/testify` library prints both the expected and the actual value of the test result, as well as providing the details about the test case (the `fmt.Sprintf` result, in our case).

Avoiding the use of Fatal in your logs

The built-in Go testing library includes different functions for logging errors, including `Error`, `Errorf`, `Fatal`, and `Fatalf`. The last two functions print the logs and interrupt the execution of the tests. Consider this test code:

```
if err := Process(tt.in); err != nil {
    t.Fatalf("Process(%v): %v, want nil", err)
}
```

The call to the `Fatalf` function interrupts the test execution. Interrupting test execution is often not the best idea because it leads to fewer tests being executed. Executing fewer tests leaves the developer with less information for the remaining failing test cases. Fixing one error and running all the tests again may be a suboptimal experience for many developers and it is often better to continue the test execution whenever possible.

The previous example can be re-written as follows:

```
if err := Process(tt.in); err != nil {
    t.Errorf("Process(%v): %v, want nil", err)
}
```

If you use this code in a loop, you can add `continue` after the `Errorf` call to proceed to the next test cases.

Making a comparison using a cmp library

Imagine that you have a test that compares the `Metadata` structure that we defined in *Chapter 2*:

```
want := &model.Metadata{ID: "123", Title: "Some title"}
id := "123"
if got := GetMetadata(ctx, "123"); got != want {
   t.Errorf("GetMetadata(%v): %v, want %v", id, got, want)
}
```

The code here would not work for structure references – in our code, the `want` variable holds a pointer to the `model.Metadata` structure, so the `!=` operator will return `true` even for structures with the same field values if these structures are created separately.

A comparison of structure pointers can be made in Go using the `reflect.DeepEqual` function:

```
if !reflect.DeepEqual(GetMetadata(ctx, "123"), want); {
   t.Errorf("GetMetadata(%v): %v, want %v", id, *got, *want)
}
```

However, the output of the test may not be easy to read. Consider that you have lots of fields inside the `Metadata` structure – if only one field is different, you will need to scan through both structures to find the difference. There is a convenient library that simplifies comparison in tests called `cmp` (`https://pkg.go.dev/github.com/google/go-cmp/cmp`).

The `cmp` library allows you to compare arbitrary Go structures in the same way as with `reflect.DeepEqual`, but it also provides human-readable output. Here's an example of using the function:

```
if diff := cmp.Diff(want, got); diff != "" {
   t.Errorf("GetMetadata(%v): mismatch (-want +got):\n%s",
tt.in, diff)
}
```

If the structures don't match, the `diff` variable will be a non-empty string, including the printable representation of the differences between them. Here's an example of this kind of output:

```
GetMetadata(123) mismatch (-want +got):
  model.Metadata{
      ID:       "123",
-     Tiitle: s"Title",
```

```
+        IPAddress: s"The Title",
     }
```

Note how the `cmp` library highlighted the differences between both structures using the – and +
prefixes. Now, it is easy to read the test output and notice the differences between the structures — this
kind of optimization will save you lots of time during debugging.

This summarizes our short collection of Go testing best practices — you can find more tips by reading
the documents mentioned in the *Further reading* section. Make sure to familiarize yourself with the
official recommendations and the comments for the `testing` package to learn how to write tests in
a conventional way and leverage all the features provided by the built-in Go testing library.

Summary

In this chapter, we covered multiple topics related to Go testing, including the common features of
the Go testing library and the basics of writing unit and integration tests for your code. You have
learned how to add tests to your microservices, optimize test execution in various cases, create test
mocks, and maximize the quality of your tests by following the best testing practices. The knowledge
you gained from reading this chapter should help you to increase the efficiency of your testing logic
and increase the reliability of your microservices.

In the next chapter, we will move to a new topic, that will cover the main aspects of service reliability
and describe various techniques for making your services resilient to various types of failures.

Further reading

- Golang test comments: `https://github.com/golang/go/wiki/TestComments`
- Golang testing package documentation: `https://pkg.go.dev/testing`
- *Using Subtests and Sub-benchmarks*: `https://go.dev/blog/subtests`

Part 3: Maintenance

This part covers some advanced topics of Go microservice development, such as reliability, observability, alerting, ownership, and security. You will learn how to handle different types of microservice-related issues, how to collect and analyze service performance data, how to set up automated service incident alerting, and how to secure communication between your microservices. The part includes lots of best practices and examples that will help to apply the newly gained knowledge to your microservices.

This contains the following chapters:

- *Chapter 10*, Reliability Overview
- *Chapter 11*, Collecting Service Telemetry Data
- *Chapter 12*, Setting up Service Alerting
- *Chapter 13*, Advanced Topics

10

Reliability Overview

We have made a long journey through all previous chapters of this book and completed the part of the book dedicated to microservice development basics. So far, you have learned how to bootstrap microservices, write tests, set up service discovery, use synchronous and asynchronous communication between your microservices, and serialize the data between them using different formats, as well as how to deploy the services and verify that their APIs work.

This chapter begins the third part of the book, dedicated to more advanced concepts of microservice development, including reliability, observability, maintainability, and scalability. In this chapter, we will cover some practical aspects of microservice development that are important for ensuring your services can operate well under many conditions, including failure scenarios, changes in network traffic, and unexpected service shutdowns.

In this chapter, we will cover various techniques and processes that can help you increase the reliability of your services. We will cover the following topics:

- Reliability basics
- Achieving reliability through automation
- Achieving reliability through development processes and culture

Let's proceed to the first section of the chapter, which will help you to understand service reliability concepts better.

Technical requirements

To complete this chapter, you need Go 1.11+ or above.

You can find the GitHub code for this chapter here:

```
https://github.com/PacktPublishing/microservices-with-go/tree/main/
Chapter10
```

Reliability basics

While implementing new applications, services, or features, engineers often focus first on meeting various system requirements, such as implementing specific application features. The initial result of such work is usually some working code that correctly performs its job, such as handling some data processing task or serving network requests as an API endpoint. We can say that such code initially performs well in isolation—the implemented code produces expected outputs for the inputs we provide.

Things usually get more complex when we add more components to the system. Let's take our movie service from *Chapter 2* and assume that its API gets used by some external service that has millions of users. Our service can be implemented perfectly fine and produce the right results for various test inputs. Still, once we get requests from an external service, we may notice various issues. One of them is called **denial of service (DoS)**—an external service can overload our service by asking to process too many requests, to the extent that our service stops serving new requests. The outcome of such an issue can vary from minor system performance degradation to service crashes due to reaching CPU, file, or memory limits.

DoS is just one of the examples of things that can go wrong in a microservice environment. Assume that you performed a fix that limits the number of incoming requests to your service, but the fix broke the services calling your API because they did not expect a sudden DoS on their requests. An alternative scenario is a change in a service API that introduces a **backward-incompatible change**. This change is incompatible with one or multiple previously released versions of callers of your service API. As a result, services calling your API could experience various negative effects, up to the point that they would be unable to process any requests.

Let's define the quality of a service that can be resilient in the face of unexpected failures as **reliability**—the quality of operating expectedly and having explicitly defined limitations. The last clause in our definition of reliability makes a big difference to its meaning—it's not enough to perform a certain function well. It is equally important to be explicit about the service's limitations and what happens when these limitations are breached.

In our movie service example, we would need to be explicit about multiple things, such as the following:

- **System throughput**: How many requests the service can process (for example, maximal requests per second)
- **Congestion policy**: How we would handle scenarios when our service is overloaded

For example, if our service can't process more than 100 simultaneous requests per service instance, we could explicitly state this in the documentation to our API and reject all extra incoming requests by returning a special error code, such as `HTTP 429 Too Many Requests`. Such indication of system limits and explicit communication of congestion issues would be a great step toward improving overall system reliability by making its behavior more deterministic and, hence, reliable.

In general, achieving a high degree of reliability is a continuous process and requires constant improvements in the following three categories:

- **Prevention**: An ability to prevent possible issues whenever possible
- **Detection**: An ability to detect possible issues as early as possible
- **Mitigation**: An ability to mitigate any issues as early as possible

Prevention, detection, and mitigation improvements can be made by performing two types of actions:

- Automating service responses to various types of failures
- Changing and improving service development processes

We will divide the rest of the chapter into two sections, describing these two types of actions. Let's proceed to the first section, covering the automation-related reliability work.

Achieving reliability through automation

In this section, we will talk about various automation techniques that can help you improve the reliability of your services.

First, let's get back to communication error handling, which we briefly covered earlier in *Chapter 5*. Having the right communication error-handling logic in place is the first step toward achieving higher reliability of your services, so we will focus on multiple aspects of error handling that are equally important in microservice development.

Communication error handling

As we discussed in *Chapter 5* of this book, when two components—such as a client and a server—communicate with each other, there are three possible resulting scenarios:

- **Successful response**: The server receives and successfully processes a request.
- **Client error**: An error occurs, and it is not caused by the server (for example, the client sends an invalid request).
- **Server error**: An error occurs, and it is caused by the server (for example, due to an application crash or an unexpected error on the server side).

From the perspective of a client, there are two different classes of errors:

- **Retriable errors**: A client may retry the original request (for example, when a server is temporarily unavailable).
- **Non-retriable errors**: A client should not retry the request (for example, when the request itself is incorrect due to failing validation).

Differentiating between retriable and non-retriable errors is the responsibility of the client. However, it is a good practice to indicate this explicitly whenever possible. For example, a server can return specific codes, indicating the types of errors (such as HTTP 404 Not Found) so that a client can recognize retriable errors and perform retries. Differentiation between client and server errors also helps to ensure that requests are not retried for non-retriable errors. It is important from the server's perspective because handling duplicate, invalid requests increases its load.

Let's illustrate how to handle retriable communication errors by implementing client request retries. Setting up automated responses to potential issues, such as communication errors, helps to make the system more resilient to transient failures, resulting in a better experience for all components in the system.

Implementing request retries

Let's illustrate how to implement request retries in microservice code. For this, let's review the metadata gRPC gateway code we implemented earlier in *Chapter 5*. The Get function includes the actual call to the metadata service:

```
resp, err := client.GetMetadata(ctx, &gen.
GetMetadataRequest{MovieId: id})
    if err != nil {
        return nil, err
    }
```

Let's now look at the implementation of the GetMetadata endpoint in the metadata service gRPC handler. The GetMetadata function includes the following code:

```
func (h *Handler) GetMetadata(ctx context.Context, req *gen.
GetMetadataRequest) (*gen.GetMetadataResponse, error) {
    if req == nil || req.MovieId == "" {
        return nil, status.Errorf(codes.InvalidArgument, "nil
req or empty id")
    }
    m, err := h.ctrl.Get(ctx, req.MovieId)
    if err != nil && errors.Is(err, metadata.ErrNotFound) {
        return nil, status.Errorf(codes.NotFound, err.Error())
    } else if err != nil {
        return nil, status.Errorf(codes.Internal, err.Error())
    }
    return &gen.GetMetadataResponse{Metadata: model.
```

```
MetadataToProto(m)}, nil
}
```

As we can see, the implementation of the `GetMetadata` endpoint includes three error cases, each having its own gRPC error code:

- `InvalidArgument`: The incoming request fails the validation.
- `NotFound`: The record with the provided identifier is not found.
- `Internal`: Internal server error.

The `InvalidArgument` and `NotFound` errors are non-retriable—there is no point in retrying requests failing validation or trying to retrieve records that are not found. `Internal` errors may indicate a wide range of issues, such as bugs in the service code, so we can't certainly state that you should perform retries on them.

There are, however, some other types of gRPC error codes that indicate potentially retriable errors. Let's list some of them:

- `DeadlineExceeded`: Indicates a problem with processing a request within the configured interval of time.
- `ResourceExhausted`: The service processing the request is exhausted. This can indicate a problem with a lack of available resources (for example, the CPU, memory, or disk reaching its limit) or the client reaching its quota for accessing the service (for example, when a service does not allow more than a certain number of parallel requests).
- `Unavailable`: The service is currently unavailable.

Let's first implement some simple retry logic inside the metadata gRPC gateway by replacing the `Get` function with the following code:

```
// Get returns movie metadata by a movie id.
func (g *Gateway) Get(ctx context.Context, id string) (*model.
Metadata, error) {
    conn, err := grpcutil.ServiceConnection(ctx, "metadata",
g.registry)
    if err != nil {
        return nil, err
    }
    defer conn.Close()
    client := gen.NewMetadataServiceClient(conn)
    var resp *model.Metadata
    const maxRetries = 5
```

```
    for i := 0; i < maxRetries; i++ {
        resp, err = client.GetMetadata(ctx, &gen.
GetMetadataRequest{MovieId: id})
        if err != nil {
            if shouldRetry(err) {
                continue
            }
            return nil, err
        }
        return model.MetadataFromProto(resp.Metadata), nil
    }
    return nil, err
}
```

Add a function that should help us to check whether a communication error is retriable:

```
func shouldRetry(err error) bool {
    e, ok := status.FromError(err)
    if !ok {
        return false
    }
    return e.Code() == codes.DeadlineExceeded || e.Code() ==
codes.ResourceExhausted || e.Code() == codes.Unavailable
}
```

Note that we also need to import two extra packages for checking for specific gRPC error codes—
google.golang.org/grpc/codes for accessing a list of error codes and google.golang.
org/grpc/status for checking whether the communication error is a valid gRPC error.

Now, our metadata gRPC gateway can perform up to five retries of requests to the metadata service.
The retry logic that we just added should help us minimize the impact of occasional errors, such as
temporary server unavailability (for example, during an unexpected outage or temporary network
issues). However, it introduces some additional challenges:

- **Extra requests to the server**: For every call to the Get function, the metadata service gRPC
  gateway now performs up to five calls instead of one for retriable errors.

- **Request bursts**: The metadata gRPC gateway performs immediate retries on errors, which will
  generate bursts of requests to the server.

The latter scenario may be especially challenging to the server due to uneven load distribution. Imagine that you are doing some work and getting some phone calls with extra tasks. If you responded to such calls and said that you were busy, you wouldn't want to get called again immediately and asked to perform the same tasks again—instead, you would want the caller to call back after some time. Similarly, immediate retries would be suboptimal to servers experiencing congestion issues, so we would need to perform additional modifications to our retry logic to introduce extra delays between the retries so that our server does not get overloaded with immediate retries.

The technique of adding extra delays between client request retries is called **backoff**. Different types of backoff are implemented by using different delay intervals between the retry requests:

- **Constant backoff**: Each retry is performed after a constant delay.

- **Exponential backoff**: Each retry is performed after a delay that is exponentially higher than the previous one.

An example of exponential backoff would be a sequence of calls where the first retry would be done after a 100 ms delay, the second one would take a 400 ms wait, and the third retry delay would be 900 ms. Exponential backoff is usually a better solution than constant, because it performs the next retry much slower than the previous ones, allowing the server to recover in case of overloading. A popular Go library at `https://github.com/cenkalti/backoff` provides an implementation of exponential and other types of backoff algorithms.

Backoff delay can also be modified by introducing small random changes to its duration. For example, the retry delay value on each step could be increased or decreased by up to 10% to better spread the load on the server. This optimization is called **jittering**. To illustrate the usefulness of jittering, assume multiple clients start calling the server simultaneously. If retries are performed with the same delays for each client, they will keep calling the server simultaneously, generating bursts of server requests. Adding pseudo-random offsets to retry delay intervals helps to distribute the load on a server more evenly, preventing possible traffic bursts from request retries.

Deadlines and timeouts

Let's now talk about another class of communication issues related to time. When a client performs a request to a server, multiple possible failures may result in either a client or a server not receiving enough data to consider the request successful. Possible failure scenarios include the following:

- The client request does not reach the server due to network issues.

- The server gets overloaded and takes longer to respond to the client.

- The server processes the request, but the response does not reach the client due to network issues.

These failures can result in longer waiting times for a client. Imagine you are sending a letter to your relative and not getting a response back. Without additional information, you would continue waiting without knowing whether the letter got lost at any step or the relative simply hasn't responded.

For synchronous requests, there is a way to improve the client experience by setting a **request timeout**—an interval after which the request is considered as failed in case of not receiving a successful response. Setting request timeouts is a good practice due to multiple reasons:

- **Elimination of unexpected waits**: If a request takes an unexpectedly long time, the client can stop it earlier and perform an optional retry.

- **Ability to estimate maximum request processing time**: When requests are performed with explicit timeouts, it is easier to calculate how long it will take until the operation returns a response or an error to the caller.

- **Ability to set longer timeouts for long-running operations**: Libraries used for performing network calls often set default request timeouts (for example, 30 seconds). Sometimes the clients want to set a higher value, knowing that the request may take longer to complete (for example, when uploading a large file to a server). Explicitly setting a higher timeout helps to prevent the situation of a request getting canceled due to exceeding the default timeout.

In Go, timeouts are usually propagated via the `context.Context` object. As we mentioned in *Chapter 1*, each I/O operation, such as a network call, accepts the `context` object as an argument, and we can set a timeout by calling the `context.WithTimeout` function, as shown in the following code snippet:

```go
func TimeoutExample(ctx context.Context, args Args) {
    const timeout = 10 * time.Second
    ctx, cancel := context.WithTimeout(ctx, timeout)
    defer cancel()
    resp, err := SomeOperation(ctx, args)
}
```

In the preceding example, we set the timeout for the `SomeOperation` function to `10` seconds, so it should not take more than 10 seconds to complete the operation.

Setting a timeout is not the only way to limit request processing time. An alternative solution to this is setting a **deadline**—the maximal time until which the request should get processed not to be considered as failed. Unlike a timeout, which is set using the `time.Duration` structure (for example, having the value of 10 seconds), a deadline indicates the exact instance of time (for example, January 1, 2074, 00:00:00). Here's an example of using a deadline for the same operation as in the previous code example:

```go
deadline := time.Parse(time.RFC3339, "2074-01-01T00:00:00Z")
ctx, cancel := context.WithDeadline(ctx, deadline)
defer cancel()
resp, err := SomeOperation(ctx, args)
```

Technically, both a timeout and a deadline help us achieve the same goal—set a time limit for a target operation. You are free to use either format, depending on your preferences.

Fallbacks

Let's now talk about another client-server communication failure scenario—when a client tries to operate and doesn't get a successful response even after a set of retries. In such a case, there are three possible options for the client:

- Return an error to the caller, if any

- Panic, in case an error is fatal to the system

- Perform an alternative backup operation, if it is possible

The last option is called a **fallback**—an alternative logic that can get executed if some operation can't be performed as expected.

Let's take our rating service as an example. In our service, we implemented the GetAggregatedRating endpoint by reading all ratings for a provided record from the rating repository. Now, let's consider a failure scenario when we can't retrieve the ratings due to some problem, such as MySQL database unavailability. Without a fallback logic, we would not be able to process an incoming request and would need to return an error to our caller.

An example of a fallback would be to use a **cache**—we could store the previously retrieved ratings in the memory of a service (for example, inside a map structure) and return them on database-read errors. The following code snippet provides an example of such a fallback logic:

```
ratings, err := c.repo.Get(ctx, recordID, recordType)
if err != nil && err == repository.ErrNotFound {
    return 0, ErrNotFound
} else if err != nil {
    log.Printf("Failed to get ratings for %v %v: %v",
recordID, recordType, err)
    log.Printf("Fallback: returning locally cached ratings
for %v %v", recordID, recordType)
    return c.getCachedRatings(recordID, recordType)
}
```

Using fallbacks is an example of **graceful degradation**—a practice of handling application failures in a way that an application still performs its operations in a limited mode. In our example, the movie service would continue processing requests for getting movie details even if the recommendation feature is unavailable, providing a limited but working functionality to its users.

When designing new services or features, ask yourself which operations could be replaced with fallbacks in case of failures. Additionally, check which features and operations are absolutely necessary and which ones can be turned off in case of any failure, such as system overload or losing a part of a system due to an outage. Also, a good practice is to emit additional useful information related to failures, such as logs and metrics, and make it explicit in the code that the fallback is intentional, as in the preceding example.

Rate limiting and throttling

As we discussed at the beginning of this chapter, there may be a situation when a microservice is overloaded and can't handle incoming requests anymore. How can we prevent or mitigate such issues?

A popular way of preventing such issues is setting a hard limit on the number of requests to be processed in parallel. Such a technique is called **rate limiting** and can be applied on multiple levels:

- **Client level**: A client limits the number of simultaneous outgoing requests.
- **Server level**: A server limits the number of simultaneous incoming requests.
- **Network/intermediate level**: The number of requests between a server and its clients is controlled by some logic or an intermediate component between them (for example, by a load balancer).

When a client or a server exceeds the configured number of requests, the result of a request would be an error that should include a special code or message, indicating that a request has been rate limited.

An example of a rate-limiting indication in the HTTP protocol is a built-in status code, `429 Too Many Requests`. When a client receives a response with such a code, it should take this into account by either reducing the call rate or waiting some time until the server can process requests again.

Client- and server-level rate limiting are often done by each service instance separately: each instance keeps track of the current number of outgoing or incoming requests. The downside of these models is the inability to configure the limits on a global-service level. If you configure each service client instance to send no more than 100 requests per second, you may still receive 100,000 simultaneous requests if there are 1,000 client instances. Such a high number of simultaneous requests could easily overload your service.

Network-level rate limiting can potentially solve this problem: if rate limiting is performed in a centralized way (for example, by a load balancer that handles requests between the services), the component performing rate limiting can keep track of the total number of requests across all service instances.

While network-level rate limiters provide more flexibility to configure the settings, they often require additional centralized components (such as load balancers). Because of this, we are going to demonstrate how to use a simpler approach, based on the client level.

There is a popular package implementing rate limiting in Go, called `golang.org/x/time/rate`. The package implements the **token bucket** algorithm—a limiting algorithm that initializes a bucket of some configured maximal size b, decrements its value by 1 on each request, and refills it at a configured rate of r elements per second. For example, for b = 100 and r = 50, the token bucket algorithm creates a bucket of size 100 and refills it at a rate of 50 per second. At any moment in time, it doesn't allow more than 100 simultaneous requests (the maximal number is controlled by the current bucket size).

Here is an example of using a token bucket-based rate limiter in Go:

```go
package main

import (
    "fmt"

    "golang.org/x/time/rate"
)

func main() {
    limit := 3
    burst := 3
    limiter := rate.NewLimiter(rate.Limit(limit), burst)
    for i := 0; i < 100; i++ {
        if limiter.Allow() {
            fmt.Println("allowed")
        } else {
            fmt.Println("not allowed")
        }
    }
}
```

This code prints allowed 3 times and then keeps printing not allowed 97 times unless it takes more than 1 second to run.

Let's illustrate how to use such a rate limiter in combination with a gRPC API handler, which we implemented in *Chapter 5*. The gRPC protocol allows us to define **interceptors**—operations that are performed on each request and can modify the gRPC server's response to it. To add a gRPC rate limiter to the movie service gRPC handler, perform the following steps:

1. Open the movie/cmd/main.go file and add the following code to its imports:

   ```
   "github.com/grpc-ecosystem/go-grpc-middleware/ratelimit"
   ```

2. Replace the line with a `grpc.NewServer` call with the following code:

```
const limit = 100
const burst = 100
l := newLimiter(100, 100)
srv := grpc.NewServer(grpc.UnaryInterceptor(ratelimit.
UnaryServerInterceptor(l)))
```

3. Then, add the following structure definition to the file:

```
type limiter struct {
    l *rate.Limiter
}

func newLimiter(limit int, burst int) *limiter {
    return &limiter{rate.NewLimiter(rate.Limit(limit),
burst)}
}

func (l *limiter) Limit() bool {
    return l.l.Allow()
}
```

Our rate limiter is using a rate-limiting gRPC server interceptor from the `github.com/grpc-ecosystem/go-grpc-middleware/ratelimit` package. Its interface is slightly different from our limiter from `golang.org/x/time/rate`, so we added a structure that links them together. Now, our gRPC server allows up to 100 requests per second and returns an error with a `codes.ResourceExhausted` special code in case the limit is exceeded. This allows us to make sure the service does not get overloaded with a sudden spike of a large number of requests—if somebody requests 1 million movie details at once from it, we are not going to make 1 million calls to our metadata service and overload its database.

Keep in mind that rate limiting is a powerful technique; however, it needs to be used with caution because setting the limit too low would make your system unnecessarily too restrictive for users by rejecting too many requests. To calculate fair rate-limiting settings for your services, you need to periodically perform benchmarking, understanding the maximum throughput of their logic.

Let's move to the next topic of automation-based reliability techniques, describing how to gracefully terminate the execution of your services.

Graceful shutdown

In this section, we are going to talk about the graceful handling of service shutdown events. Service shutdowns can be triggered by multiple events:

- Manual interruption of execution (for example, when a user types *Ctrl + C/Cmd + C* in a terminal that runs the service process, and the process receives a `SIGINT` signal from the operating system)

- Termination of execution by the operating system (for example, by `SIGTERM` or `SIGKILL` signals)

- Panic in service code

Generally, sudden termination of the execution of a service may result in the following negative consequences:

- **Dropped requests**: Incoming API requests may be dropped before they get fully processed, resulting in errors for the callers of the service.

- **Connection issues**: Service network connections may not be properly closed during a shutdown, resulting in multiple negative effects. For example, not closing a database connection may result in a situation called a **connection leak**, when the database keeps the connection allocated to the service instead of allowing it to be reused by another instance.

To prevent these issues, you need to ensure that your service shuts down gracefully by performing a set of operations that minimize any negative consequences for the service and its components. Performing a **graceful shutdown**, the service would run some extra logic before the termination, such as the following:

- Completing as many unfinished operations, such as unprocessed requests, as possible

- Closing all open network connections and yielding any shared resources, such as network sockets

Graceful shutdown logic for Go services is usually implemented in the following way:

1. The service subscribes to shutdown events by calling a `Notify` function of an `os/signal` package.

2. When a service receives a `SIGINT` or `SIGTERM` event from the operating system, indicating that the service is about to be terminated, it performs a set of required operations for closing all open connections and completing all pending tasks.

3. Once all operations are completed, the service finishes the execution.

Here is a code example that you can add to the `main` function of any Go service, such as the ones that we implemented in *Chapter 2*:

```
sigChan := make(chan os.Signal, 1)
signal.Notify(sigChan, os.Interrupt, syscall.SIGTERM)
var wg sync.WaitGroup
wg.Add(1)
go func() {
    defer wg.Done()
    s := <-sigChan
    log.Printf("Received signal %v, attempting graceful
shutdown", s)
    // Graceful shutdown logic.
}()
wg.Wait()
```

There is also a way to gracefully handle panics in Go code by using the built-in `recover` function. The following code snippet demonstrates how to handle a panic inside the `main` function and execute any custom logic, such as closing any open connections:

```
func main() {
    defer func() {
        if err := recover(); err != nil {
            log.Printf("Panic occurred, attempting graceful
shutdown")
            // Graceful shutdown logic.
        }
    }()
    panic("panic example")
}
```

In our code, we check whether there is a service panic by calling the `recover` function and checking whether it returns a non-nil error. In case of a panic, we can perform any additional operations, such as saving any unsaved data or terminating any open connections.

To gracefully terminate the execution of a Go gRPC server, you need to call the `GracefulStop` function instead of `Stop`. Unlike the `Stop` function, `GracefulStop` would wait until all requests are processed, helping to reduce the negative impact of the shutdown on the clients.

If you have some long-running components, such as Kafka consumers or any background goroutines executing long-running tasks, you can communicate the service termination signal using the built-in context.Context structure. The context.Context structure provides a feature called **context cancellation**—an ability to notify different components about the cancellation of an execution by sending a specific event through the channel associated with the context.

Let's update our rating service code to illustrate how to implement context cancellation and a graceful shutdown of a gRPC server:

4. Open the main.go file of the rating service and find the line that performs a call to the context.Background() function. Replace it with the following code:

ctx, cancel := context.WithCancel(context.Background())

Our code creates an instance of a context and the cancel function, which we will be calling on service shutdown to notify our components, such as the service registry, about upcoming termination.

5. Immediately before the call to the srv.Serve function, add the following code:

```
sigChan := make(chan os.Signal, 1)
signal.Notify(sigChan, os.Interrupt, syscall.SIGTERM)
var wg sync.WaitGroup
wg.Add(1)
go func() {
    defer wg.Done()
    s := <-sigChan
    cancel()
    log.Printf("Received signal %v, attempting graceful
shutdown", s)
    srv.GracefulStop()
    log.Println("Gracefully stopped the gRPC server")
}()
```

In our code, we let the rating service listen for process interruption and termination signals and start the background goroutine, which keeps listening for the relevant notifications. Once it receives either signal, it calls the cancel function that we obtained in the previous step. The result of calling this function would be a notification that would be sent to the components initialized with our context, such as the service registry.

6. Let's add the final touch by adding the following line to the end of our main function:

```
wg.Wait()
```

Let's now test the code that we just implemented. Run the rating service and then terminate it by pressing *Ctrl + C*/*Cmd + C* (depending on your OS). You should see the following messages:

```
2022/10/13 08:55:05 Received signal interrupt, attempting
graceful shutdown
2022/10/13 08:55:05 Gracefully stopped the gRPC server
```

Communication of termination and interruption events is a common practice in Go microservice development and is an elegant way of implementing graceful shutdown logic. When designing and implementing your services, think in advance about possible resources that need to be closed or de-initialized upon the service termination, such as any network clients and connections. A graceful shutdown logic can prevent the negative effects of sudden service termination. It can also reduce the number of possible errors in your services and improve your operating experience.

At this point, we have reviewed some automation techniques to improve the reliability of our services and reduce the symptoms of various failure scenarios. Now, we can proceed to the next section of the chapter, covering another aspect of reliability work related to development processes and culture. Improvements to your development processes are essential to achieving high reliability in the long term, and the section should be useful to you by providing some valuable tips and ideas that you can utilize in microservice development.

Achieving reliability through development processes and culture

In this section, we are going to describe some techniques for achieving higher service reliability based on changes in the development processes and culture. You will learn how to establish the processes for improving and reviewing your service reliability, how to learn from any service-related issues and incidents efficiently, and how to measure your service reliability. We will cover the processes and practices that are widely used across the industry, outlining the most important ideas from each one. The section is going to be more theoretical than the previous one; however, it should be equally useful.

First, we are going to provide an overview of the on-call process essential for setting up a mechanism for monitoring issues with your services.

On-call process

When your services start handling production traffic or start serving user requests, one of your first reliability goals should be to detect any issues, or incidents, as early as possible. Efficient detection should be automatic—a program will always be much more efficient than a human in detecting most issues. Each automatic detection should notify one or more engineers about the incident so that the engineers can perform work in order to mitigate an incident.

The process for establishing such a mechanism for notifying engineers about service incidents is called **on-call**. This process helps ensure that at any moment in time, service incidents are acknowledged and addressed by the engineers responsible for the service.

The main ideas behind the on-call process are the following:

- Engineers can get grouped into **on-call rotations**. Each engineer participating in the on-call process repeatedly gets assigned a continuous *shift* (often taking 1 week), during which they take responsibility for periodically handling notifications regarding service-level incidents.

- On-call rotation can have an *escalation policy*—a process of escalating incidents in case they remain unresolved. First, an incident gets reported to the *primary* on-call engineer of the rotation. If the primary engineer is unavailable, the incident gets reported to the *secondary* engineer, and so on.

- There can be a *shadow* role, commonly assigned to new engineers. This role does not require any response to the incident, but it can be used for getting familiar with the on-call process and subscribing to real-time incident notifications.

- Each incident triggers one or multiple notifications, notifying the on-call engineers about the issue. Each notification must be acknowledged by the responsible on-call engineer unless the incident self-resolves (for example, if a service stops receiving too many requests and starts operating normally).

- You can also set up an *escalation policy* for a rotation—a mechanism for escalating the incident notifications if the responsible on-call engineers don't acknowledge them within the configured time. Usually, escalation policy follows the reporting chain of the engineering hierarchy—if no engineer acknowledges the incident, the incident first triggers a notification to the closest engineering manager, then to the person the manager is reporting to, and so on until it reaches the highest level (this can even be a CTO at some companies).

Having an on-call process is common to most technology companies and teams, and the on-call process is pretty similar in most companies across the industry. Some popular solutions provide mechanisms for triggering various types of notifications, such as SMS, emails, and even phone calls. You can also configure on-call rotations and assign them to different services. One of the most popular solutions to on-call management is **PagerDuty**—a platform providing a set of tools for automating on-call operations, as well as integrations with hundreds of services, including Slack, Zoom, and many more. PagerDuty provides all the features we listed earlier, allowing engineers to configure on-call rotations for their services and notifying them about incidents in different ways. Additionally, it provides an API that can be used for both accessing the incident data and triggering new incidents from the code.

We are not going to dive into the details of PagerDuty's features and integrations in this chapter—I suggest you check the official PagerDuty documentation on their website, `https://developer.pagerduty.com/docs`. I also suggest you read *Chapter 12* before establishing an on-call process for your services. It will help you to learn more about possible incident detection mechanisms and tools you can utilize in your projects.

Let's discuss the common challenges of establishing an on-call process in a microservice environment:

- **Rotation ownership**: Different services may be maintained by different teams, so there may be multiple on-call rotations inside a single company. A good practice is to have an explicit mapping between each production service and the associated on-call rotation so that it is clear which rotation each incident should be reported to. In *Chapter 13*, we will cover the ownership aspect of this.

- **Cross-service issues**: Some issues, such as database or network failures, can span multiple services, so it becomes important to have some centralized team(s) that will be able to help with any issues crossing the boundaries of individual services.

Some companies may have thousands of microservices, so centralized incident response teams become crucial. For example, Uber has a dedicated team of engineers called *Ring0* that is able to address any widespread incidents and coordinate the mitigation of issues that span multiple teams. Having such a team helps to dramatically reduce incident mitigation time.

To better understand what happens next after incidents are detected and acknowledged by the engineers, we are going to move now to the next topic: incident management.

Incident management

Once incidents are detected and acknowledged by the engineers, there are two other types of work necessary for improving the service or system reliability—mitigation and prevention. Mitigation is required for resolving an open issue unless it gets resolved by itself or due to some external changes (for example, an external API getting fixed by the owning team). Prevention work is useful for ensuring the issue does not happen again. Without a proper prevention response to the incident, you may keep fixing the same issue over and over again, spending your time and affecting the experience of your system's users.

To make the incident mitigation process quick and efficient, especially in a large team where engineers may have different levels of understanding of the system, there should be enough documentation describing which actions to perform in case of an incident. Such documentation is called a **runbook** and should be prepared for as many types of detectable incidents as possible. Whenever an on-call engineer gets an incident notification, it should be clear from the runbook which steps to take to mitigate it.

A good runbook should be short and concise and provide clear actionable steps that are easy to understand for any engineer. Let's take this example:

```
rating_service_fd_limit_reached:
  mitigation: Restart the service
```

If the incident mitigation requires further investigation, include any useful links, such as links to the relevant application logs and dashboards. You should aim for the lowest possible incident mitigation time—also called **time to repair** (**TTR**)—to increase the availability of your service and improve its overall health.

Once the incident is mitigated, focus on prevention work to ensure you take all actions to eliminate its causes, as well as to improve detection and mitigation mechanisms, if needed. Multiple companies across the industry use the process of writing documents called **incident postmortems** to organize the learnings around incidents and make sure each incident involves enough work related to its future prevention. An incident postmortem generally consists of the following data:

- Incident title and summary
- Authors
- When and how the incident was detected and mitigated
- Incident context, in the form of a text or a set of diagrams that can help to understand it
- Root cause
- Incident impact
- Incident timeline
- Lessons learned
- Action items

A great example of a postmortem document is provided in the famous Google *Site Reliability Engineering (SRE)* book, and you can get familiar with it at the following link: `https://sre.google/sre-book/example-postmortem/`.

To get to the root cause of the incident, you can use a technique called **Five whys**. The idea of the technique is to keep asking what caused the previously mentioned problem until the root cause is found. Let's take the following **root cause analysis** (**RCA**) as an example to understand the technique:

Incident: Rating service returns internal errors to its API callers

Root cause analysis:

1. The rating service started returning internal errors to its API callers due to the rating database's unavailability.
2. The rating database became unavailable because of an unexpectedly high request load to it.
3. The unexpectedly high request load to the rating service was caused by an application bug in the movie service.

In this example, we kept finding the underlying cause of each previous issue by using the Five whys technique, until we got to the root cause of the incident in just three steps. The technique is very powerful and easy to use, and it can help you get to the root cause of even complex issues quite quickly.

Make sure you include and track action items for your incidents. Capturing the incident details and identifying the causes isn't enough for making sure incidents are prevented. Prioritizing the action items helps ensure that the most critical ones get addressed as early as possible.

Now, let's move to the next reliability process based on periodic testing of your possible service failure scenarios.

Reliability drills

As many system administrators know, it is not enough to have backups of your data to guarantee its durability. You also need to ensure that you can restore the data from the backups in case of any failure. The same principle applies to any part of your service infrastructure—to know that your services are resilient to particular failures, you need to perform periodic exercises, called **drills**.

You can perform many possible types of drills. As in the example with the database backups, if you have any persistent data stored in a database, you can periodically test the ability to back up and restore the data, verifying that your services are tolerable to database availability issues. Another example would be network drills. You can simulate network issues, such as connectivity loss, by updating service routing configuration or any other network settings to check how your services behave in case of network unavailability.

There are multiple benefits of performing reliability drills:

- **Detect unexpected service failures**: By performing failure drills, you can detect some unexpected service errors and panics, that don't happen in the regular mode. Such issues will present themselves in a controlled environment, where engineers are ready to stop the drill at any moment and address the detected errors as early as possible.

- **Detect unexpected service dependencies**: Reliability drills often uncover unexpected dependencies between the services, such as **transitive dependencies** (service A depends on service B, which depends on service C) or even **circular dependencies** (two services require each other in order to operate).

- **Ability to mitigate future incidents quicker**: By knowing how the services operate in case of a failure and how they resolve related issues, you invest in improving future incident mitigation.

Drills are often performed as **planned incidents**—incidents that get announced in advance and follow the regular incident management process, including the work on the postmortem document. The drill postmortem document should include the same items as a regular incident, with a focus on improving the mitigation and prevention experience. Additionally, engineers should focus on reviewing and updating the service runbooks, making sure that the incident mitigation instructions are accurate and up to date.

At this point, we have discussed the most important service reliability techniques. There are many more interesting topics to cover that are related to service reliability—some of them, related to incident detection, we are going to cover in *Chapter 12* of the book. If you are interested in the topic, I strongly encourage you to read the Google *Site Reliability Engineering (SRE)* book, which provides a comprehensive guide to various reliability-related techniques. You can find the online version of the book by going to the following link: `https://sre.google/sre-book/table-of-contents`. The practices that are described in the book are applicable to any microservice, so you can always use it as a reference while working on building any type of system.

Summary

In this chapter, we covered the topic of reliability, describing a set of techniques and practices that can help you to make your microservices more resilient to various types of failures. You have learned some useful techniques for automating error responses of your services and reducing the negative impact of various types of issues, such as service overloading and unexpected service shutdowns.

In the final part of the chapter, we discussed various reliability techniques based on changes in engineering processes and culture, such as introducing the on-call and incident management processes, as well as performing periodic reliability drills. The knowledge that you gained from reading this chapter should help you to establish a solid foundation for writing reliable microservices.

In the next chapter, we are going to continue our journey into the reliability topic and focus on collecting service telemetry data, such as logs, metrics, and traces. Service telemetry data is the primary instrument for setting up service incident detection, and we will illustrate how to work with each type of telemetry data in your microservice code.

Further reading

If you'd like to learn more, refer to the following resources:

- *Timeouts, retries, and backoff with jitter*: `https://aws.amazon.com/builders-library/timeouts-retries-and-backoff-with-jitter`

- *Token bucket rate-limiting algorithm*: `https://en.wikipedia.org/wiki/Token_bucket`

- *PagerDuty documentation*: `https://developer.pagerduty.com/docs`

- *Incident postmortems*: `https://www.pagerduty.com/resources/learn/incident-postmortem/`

- *Google Site Reliability Engineering (SRE)* website: `https://sre.google/`

- *Google Site Reliability Engineering (SRE)* book: `https://sre.google/sre-book/table-of-contents`

11

Collecting Service Telemetry Data

In the previous chapter, we explored the topic of service reliability and described various techniques for making your services more resilient to different types of errors. You learned that reliability-related work consists of making constant improvements in incident detection, mitigation, and prevention techniques.

In this chapter, we are going to take a closer look at various types of service performance data, which is essential for setting up service health monitoring and debugging and automating service incident detection. You will learn how to collect service logs, metrics, and traces, and how to visualize and debug communication between your microservices using the distributed tracing technique.

We will cover the following topics:

- Telemetry overview
- Collecting service logs
- Collecting service metrics
- Collecting service traces

Now, let's proceed to the overview of all the techniques that we are going to describe in this chapter.

Technical requirements

To complete this chapter, you will need Go 1.11+ or above. You will also need the following tools:

- **grpcurl**: `https://github.com/fullstorydev/grpcurl`
- **Jaeger**: `https://www.jaegertracing.io/`

You can find the GitHub code for this chapter here: `https://github.com/PacktPublishing/microservices-with-go/tree/main/Chapter11`.

Telemetry overview

In the introduction to this chapter, we mentioned that there are different types of service performance data, all of which are essential for service health monitoring and troubleshooting. These types of data are called **telemetry data** and include the following:

- **Logs**: Messages recorded by your services that provide insights into the operations they perform or errors they encounter

- **Metrics**: Performance data produced by your services, such as the number of registered users, API request error rate, or percentage of free disk space

- **Traces**: Data that shows how your services perform various operations, such as API requests, which other services they call, which internal operations they perform, and how long these operations take

Telemetry data is *immutable*: it captures events that have already happened to the service and provides the results of various measurements, such as service API response latency. When different types of telemetry data are combined, they become a powerful source of information about service behavior.

In this chapter, we are going to describe how to collect service telemetry data to monitor the health of services. There are two types of service health and performance monitoring:

- **White-box monitoring**: Monitoring services while having access to different types of internally produced data. For example, you can monitor a server's CPU utilization by viewing it in the system monitoring application.

- **Black-box monitoring**: Monitoring services using only externally available data and indicators. In this case, you don't know or have access to data related to their structure or internal behavior. For example, if a service has a publicly available health check API, an external system can monitor its health by calling that API without having access to internal service data.

Both types of monitoring are powered by collecting and continuously analyzing service performance data. In general, the more types of data you collect from your application, the more opportunities you get for extracting various types of information about its health and behavior. Let's list some of the ways you can use the information about your service performance:

- **Trend analysis**: Detect any trends in your service performance data:

  - Is your service health getting better or worse over time?

  - How does your API success rate change?

  - How many new users are you getting compared to the previous day/month/year?

- **Semantic graph capturing**: Capture data on how your services communicate with each other and with any other components, such as databases, external APIs, and message brokers.

- **Anomaly detection**: Automatically detect anomalies in your service behavior, such as sudden drops in API requests.

- **Event correlation**: Detect relationships between various types of events, such as unsuccessful deployments and service panics.

While observability opens lots of opportunities, it comes with the following challenges:

- **Collecting large datasets**: Real-time performance data often takes lots of space to store, especially if you have lots of services or if your services produce lots of data.

- **Need for specific tooling**: To collect, process, and visualize different types of data, such as logs, metrics, and traces, you need some extra tools. These tools often come at a price.

- **Complex setup**: Observability tooling and infrastructure are often difficult to configure. To access all data coming from multiple services, you need to set up the proper data collection, aggregation, data retention policies, and many more strategies.

We are going to describe how to work with each type of telemetry data that you can collect in your microservices. For each type of data, we will provide some usage examples and describe the common ways of setting up the tooling for working with it. First, let's proceed to look at service log collection.

Collecting service logs

Logging is a technique that involves collecting real-time application performance data in the form of a time-ordered set of messages called a **log**. Here is an example of a service log:

```
2022/06/06 23:00:00 Service started
2022/06/06 23:00:01 Connecting to the database
2022/06/06 23:00:11 Unable to connect to the database: timeout
error
```

Logs can help us understand what was happening in the application at a particular moment in time. As you can see in the preceding example, the service started at 11 P.M. and began connecting to the database a second later, finally logging a timeout error 10 seconds later.

Logs can provide lots of valuable insights about the component that emitted them, such as the following:

- **Order of operations**: Logs can help us understand the logical sequence of operations performed by a service by showing when each operation took place.

- **Failed operations**: One of the most useful applications of logs is the ability to see the list of errors recorded by a service.

- **Panics**: If a service experiences an unexpected shutdown due to panic, a log can provide the relevant information, helping troubleshoot the issue.

- **Debugging information**: Developers can log various types of additional information, such as request parameters or headers, that can help when debugging various issues.

- **Warnings**: Logs can indicate various system-level warnings, such as low disk space, that can be used as notification mechanisms for preventing various types of errors.

We used logs in the services that we created in *Chapter 2* – our services have been logging some important status messages via the built-in log library. Here's an example:

```
log.Printf("Starting the metadata service on port %d", port)
```

The built-in log library provides functionality for logging arbitrary text messages and panics. The output of the preceding operation would be as follows:

```
2022/07/01 13:05:21 Starting the metadata service on port 8081
```

By default, the log library records all logs to the `stdout` stream associated with the current process. But it is possible to set the output destination by calling the `SetOutput` function. This way, you can write your logs to files or send them over the network.

Two types of functions are provided by the log library that can be used if a service experiences an unexpected or non-recoverable error:

- `fatal`: Functions with this prefix immediately stop the execution of the process after logging the message.

- `panic`: After logging the message, they call the Go `panic` function, writing the output of the associated error. The following is an example output of calling a `panic` function:

```
2022/11/10 23:00:00 network unavailable
panic: network unavailable
```

While the built-in log library provides a simple way of logging arbitrary text messages, it lacks some useful functionality that makes it easier to collect and process the service logs. Among the missing features is the ability to log events in popular serialization formats, such as JSON, which would simplify how message data is parsed. Another issue is that it lacks `Error` and `Errorf` functions, which could be used for explicitly logging errors. Since the built-in logging library only provides a `Print` function, it's unclear by default whether the logged message indicates an error, a warning, or neither.

Yet, the biggest missing piece in the built-in log library is the ability to perform structured logging. **Structured logging** is a technique that involves collecting log messages in the form of serialized structures, such as JSON records. A distinct feature of such structures, compared to arbitrary text strings, is that they can contain additional metadata in the form of fields – key-value records. This allows the service to represent the message metadata as any supported type, such as a number, string, or serialized record.

The following snippet includes an example of a JSON-encoded log structure:

```
{"level":"info", "time":"2022-09-04T20:10:10+1:00","message":"S
ervice started", "service":"metadata"}
```

As you may have noticed, in addition to using JSON as the output format, there are two additional features of the preceding log format:

- **Log level**: There is a field called `level` that specifies the type of a log message.
- **Additional message fields**: Our example includes a field called `service`, which is used to indicate the service that emitted the message.

The output format described earlier allows us to decode the log messages much easier. It also helps us interpret their contents based on the log level and the additional message fields. Additional metadata can also be used for searching: for example, we can search for messages that have a particular `service` field value.

Now, let's focus on log-level metadata from the previous example. First, let's review some of the common log levels:

- **Info**: Informational messages that do not indicate any error. An example of such a message is a log record indicating that the service successfully connected to the database.
- **Error**: Messages indicating errors, such as network timeouts.
- **Warning**: Messages indicating some potential issues, such as too many open files.
- **Fatal**: Messages indicating critical or non-recoverable errors, such as insufficient memory, that make executing the service further impossible.
- **Debug**: Messages that provide some additional context to the developers that can help troubleshoot various issues or get some additional insights for application performance. Collecting `Debug` messages is usually disabled by default, as they often generate a large amount of data.

Log levels also help us interpret log messages. Consider the following unstructured message produced by the built-in log library, which does not include any level information:

```
2022/06/06 23:00:00 Connection terminated
```

Can you tell whether it's a regular informational message indicating the regular behavior (for example, the service intentionally terminated a connection after performing some work), a warning, or an error? If this is an error, is it critical or not? Without the log level providing additional context, it is difficult to interpret this message.

Another advantage of explicitly using log levels is the ability to enable or disable that ability to log specific types of levels. For example, logging `Debug` messages can be disabled under normal service conditions and enabled during troubleshooting. `Debug` messages often include much more information than regular ones, requiring more disk space and making it harder to navigate the other types of logs. Different logging libraries let us enable or disable specific levels, such as `Debug` or even `Info`, leaving only logs indicating warnings, errors, fatal errors, and panics.

Let's review some popular Go logging libraries and focus on choosing the one that we would use in our microservices.

Choosing the logging library

In this section, we will describe some of the existing Go logging libraries and review their features. This section should help you choose the logging library that you will use in your microservices.

First, let's list the features that we would like to get from the logging library:

- **Structured logging**: Supports logging structured messages that may include additional fields in a key-value format.

- **Fast performance**: Writing log messages should not have a noticeable impact on service performance.

- **Log level support**: Enforce inclusion of a log level in message metadata.

The following is an additional feature that would be nice to have:

- **Support formatting**: We can write formatted strings in a `Printf`-like format (for example, support an `Errorf` function to log a formatted error).

Now, let's review some of the most popular Go logging libraries. When evaluating library performance, we will be using the logging library benchmark data: `https://github.com/uber-go/zap#performance`.

The list of popular Go logging libraries includes the following:

- **Built-in Go log package** (`https://pkg.go.dev/log`):

  - Officially supported by the Go development team, included in the Go SDK

  - Does not support structured logging and does not have built-in support for log levels

- **zap** (`https://github.com/uber-go/zap`):

  - Fastest performance among all logging libraries reviewed

  - Feature-rich and supports a fast minimalistic logger, as well as a slightly slower one with additional features

- **zerolog** (`https://github.com/rs/zerolog`):
  - Fast performance
  - Simple and elegant API

- **go-kit/log** (`https://github.com/go-kit/log`):
  - Part of a larger `go-kit` toolkit for microservice development
  - Slightly slower than `zerolog` and `zap`, but faster than the other logging libraries

- **apex/log** (`https://github.com/apex/log`):
  - Has built-in support for various log storages, such as Elasticsearch, Graylog, and AWS Kinesis

- **log15** (`https://github.com/inconshreveable/log15`):
  - Feature-rich logging toolkit
  - Much slower than the other log libraries reviewed

The preceding list provides some high-level details about some of the popular Go logging libraries to help you choose the right one for your services. All libraries, except the built-in `log` package, provide the features that we need, including structured logging and log levels. Now, the question is, how do we select the best one among them?

My personal opinion is that the `zap` library provides the most flexible and yet most performant solution to service logging problems. It allows us to use two separate loggers, called `Logger` and `SugaredLogger`. `Logger` can be used in high-performance applications, while `SugaredLogger` can be used when you need some extra features; we will review these features in the next section.

Using logging features

Let's start practicing and demonstrate how to use some features of the `zap` logging library that we picked in the previous section. First, let's start with the basics and illustrate how to log a simple message that has the `Info` level and an additional metadata field called `serviceName`. The complete Go code for this example is as follows:

```
package main

import "go.uber.org/zap"

func main() {
    logger, _ := zap.NewProduction()
    logger.Info("Started the service", zap.
```

```
String("serviceName", "metadata"))
}
```

We initialize the `logger` variable by calling the `zap.NewProduction` function, which returns a production-configured logger. This logger omits debug messages, uses JSON as the output format, and includes stack traces in the logs. Then, we create a structured log message by including a `serviceName` field by using the `zap.String` function, which can be used to log string data.

The output of the preceding example is as follows:

```
{"level":"info","ts":1257894000,"caller":"sandbox1575103092/
prog.go:11","msg":"Started the
service","serviceName":"metadata"}
```

The `zap` library offers support for other types of Go primitives, such as `int`, `long`, `bool`, and many more. Corresponding functions for creating log field names follow the same naming format, such as `Int`, `Long`, and `Bool`. Additionally, `zap` includes a set of functions for the other built-in Go types, such as `time.Duration`. The following code shows an example of a `time.Duration` field:

```
logger.Info("Request timed out", zap.Duration("timeout",
10*time.Second))
```

Let's illustrate how to log arbitrary objects, such as structures. In *Chapter 2*, we defined the `Metadata` structure:

```
// Metadata defines the movie metadata.
type Metadata struct {
    ID          string `json:"id"`
    Title       string `json:"title"`
    Description string `json:"description"`
    Director    string `json:"director"`
}
```

Let's assume that we want to log the entire structure for debugging purposes. One way of doing so is to use the `zap.Stringer` field. This field allows us to log any structure or interface with the `String()` function. We can define a `String` function for our `Metadata` structure as follows:

```
func (m *Metadata) String() string {
    return fmt.Sprintf("Metadata{id=%s, title=%s,
description=%s, director=%s}", m.ID, m.Title, m.Description,
m.Director)
}
```

Now, we can log the `Metadata` structure as a log field:

```
logger.Debug("Retrieved movie metadata", zap.
Stringer("metadata", metadata))
```

The output would look as follows:

```
{"level":"debug","msg":"Retrieved movie
metadata","metadata":"Metadata{id=id, title=title,
description=description, director=director}"}
```

Now, let's illustrate one more useful technique of using the `zap` library. If you want to include the same fields in multiple messages, you can re-initialize the logger by using the `With` function, as illustrated in the following example:

```
logger = logger.With(zap.String("endpoint", "PutRating"), zap.
String("ratingId", ratingID))
logger.Debug("Received a PutRating request")
// endpoint logic
logger.Debug("Processed a PutRating request")
```

The results of both calls to the `Debug` function will now include both the `endpoint` and `ratingId` fields.

You can also use this technique when you create new service components in your code. In the following example, we are creating a sub-logger inside the `New` function:

```
func New(logger *zap.Logger, ctrl *rating.Controller) *Handler{
    return &Handler{logger.With("component":
"ratingController"), ctrl}
}
```

This way, the newly created instance of a `Handler` structure will be initialized with a logger that includes the `component` field with the `ratingController` value in each message.

Now that we have covered some of the primary service logging use cases, let's discuss how to store logs in a microservice environment.

Storing microservice logs

By default, logs of each service instance are written to the output stream of the process running it. This mechanism of log collection allows us to monitor service operations by continuously reading the data from the associated stream (`stdout` in most cases) on a host running a service instance. However, without any additional software, the log data would not be persisted, so you would not be able to read your previously recorded logs after a service restart or a sudden crash.

Various software solutions allow us to store and query the log data in a multi-service environment. They help solve multiple other problems:

- **Distributed log collection**: If you have multiple services running on different hosts, you must collect the service logs on each host independently and send them for further aggregation.

- **Centralized log storage**: To be able to query the data that's emitted by different services, you need to store it in a centralized way – all logs across all services should be accessible during the query execution.

- **Data retention**: Logging data usually takes a lot of disk space, and it often becomes too expensive to store it indefinitely for all your services. To solve this problem, you need to establish the right data retention policies for your services that will allow you to configure how long you can store the data for each one.

- **Efficient indexing**: To be able to quickly query your logging data, the logs need to be indexed and stored efficiently. Modern indexing software can help you query terabytes of log data in under 10 milliseconds.

Different tools help facilitate such log operations, such as Elasticsearch and Graylog. Let's briefly review Elasticsearch to provide an example of an end-to-end log management solution.

Elasticsearch is a popular open source search engine that was created in 2010 and quickly gained popularity as a scalable system for indexing and querying different types of structured data. While the primary use case of Elasticsearch is a full-text search, it can be efficiently used for storing and querying various types of structured data, such as service logs. Elasticsearch is also a part of the toolkit called the **Elastic Stack**, also called **ELK**, which includes some other systems:

- **Logstash**: A data processing pipeline that can collect, aggregate, and transform various types of data, such as service logs

- **Kibana**: A user interface for accessing the data in Elasticsearch, providing convenient visualization and querying features

The log collection pipeline in the Elastic Stack looks like this:

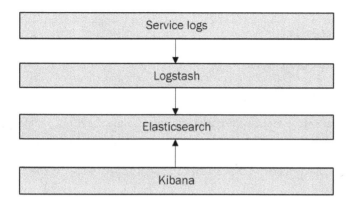

Figure 11.1 – Logging pipeline in the Elastic Stack

In this flow chart, service logs are collected by Logstash and sent to Elasticsearch for indexing and storing. Then, users can access the logs and other data indexed in Elasticsearch using the Kibana interface.

One of the key advantages of the Elastic Stack is that most of its tools are available for free and are open source. It is well-maintained and extremely popular in the developer community, making it easier to search for relevant documentation, get additional support, or find some additional tooling. It also has a set of libraries for all popular languages, allowing us to perform various types of queries and API calls to all components of the pipeline.

The Go library for using the Elasticsearch API is called `go-elasticsearch` and can be found on GitHub at `https://github.com/elastic/go-elasticsearch`.

We are not going to cover the Elastic Stack in detail as it's outside of the scope of this chapter, but you can get more familiar with the Elastic Stack by reading its official documentation (`https://www.elastic.co/guide/index.html`).

Having covered some high-level details regarding some popular logging software, let's move on to the next topic: describing the best practices of logging.

Logging best practices

So far, we have covered the most important aspects of logging and described how to choose a logging library, as well as how to establish the logging infrastructure for collecting and analyzing data. Let's describe some of the best practices for logging service data:

- Avoid using interpolated strings.
- Standardize your log messages.

- Periodically review your log data.
- Set up appropriate log retention.
- Identify the message source in logs.

Let's now cover each practice in detail.

Avoid using interpolated strings

One of the top logging anti-patterns is the usage of **interpolated strings** – messages that embed metadata inside text fields. Let's take the following snippet of code as an example:

```
logger.Infof("User %s successfully registered", userID)
```

The problem with this code is that it merges two types of data into a single text message: an operation name (user registration) and a user identifier. Such messages make it harder to search and process log metadata: each time you need to extract `userID` from a log message, you would need to parse a string that contains it.

Let's update our example by following the structured logging approach, where we log additional metadata as message fields:

```
logger.Infof("User successfully registered", zap.
String("userId", userID))
```

The updated version makes a big difference when you want to query your data. Now, you can query all log events that have `User successfully registered` text messages and easily access all user identifiers associated with them. Avoiding interpolated messages helps keep your log data easy to query and parse, simplifying all operations with it.

Standardize your log messages

In this section, we covered the benefits of log centralization and the advantages of querying the data across multiple services. But I would like to emphasize how it is important to standardize the format of log messages in a microservice environment. Sometimes, it is useful to execute log queries that span multiple services, API endpoint handlers, or other components. For example, you may need to perform the following types of queries on your log data:

- Get the distribution of timeout errors across all services.
- Get the daily count of errors for each API endpoint.
- Get distinct error messages across all database repositories.

If your services log the data using different field names, you will not be able to easily gather such data using a common query function. On the opposite side, establishing the common field names helps ensure the log messages follow the same naming convention, simplifying any queries you write.

To make sure the logs are emitted in the same way across all services and all components, you may follow these tips:

- Create a shared package that includes log field names as constants; take the following example:

```
package logging

const (
  FieldService  = "service"
  FieldEndpoint = "endpoint"
...
)
```

- To avoid forgetting to include some important field inside a certain structure, function, or set of functions, re-initialize the logger by setting the field as early as possible; take the following example:

```
func (h *Handler) PutRating(ctx context.Context, req
*PutRatingRequest) (*PutRatingResponse, error) {
    logger := h.logger.With(logging.FieldEndpoint,
"putRating")
    // Now we can make sure the endpoint field is set
across all handler logic.
...
}
```

- Additionally, ensure that the root logger of your service is setting the service name so that all your service components will automatically collect this field by default:

```
func main() {
    logger, _ := zap.NewProduction()
    logger = logger.With(logging.FieldService, "rating")
    // Pass the initialized logger to all service
components.
```

The tips that we just provided should help you standardize the usage of common fields across all your service components, making it easier to query the logged data and aggregate it in different ways.

Periodically review your log data

Once you start collecting your service logs, it is important to periodically review them. Look out for the following cases:

- **Make sure there is no PII data in logs**: **Personally identifiable information** (PII), such as full names and SSNs, falls under many regulations and generally must not be stored in logs. Make sure that no component, such as an API handler or a repository component, emits any such data, even for debugging.

- **Check that your service doesn't emit extra debug data**: Sometimes, developers log some additional data, such as request fields, to debug various issues. Check that no service is continuously emitting too many debug messages during a prolonged period, polluting the logs and using too much disk space.

Set up appropriate log retention

Log data often takes a lot of space to store. If it keeps growing in size without any additional actions being taken, you may end up using all your disk space and having to urgently clean up the old records. To prevent this, various log storage solutions allow you to configure **retention policies** for your data. For example, you can configure your log storage to keep the logs for some services for up to a few years, while limiting some other services to just a few days, depending on the requirements. Additionally, you can set some size constraints so that the logs of your services don't exceed a predefined size threshold.

Ensure you set retention policies for all types of your logs, avoiding situations when you need to clean up unneeded log records manually.

Identify the message source in logs

Imagine that you are viewing your system logs and notice the following error event:

```
{"level":"error", "time":"2022-09-04T20:10:10+1:00","message":"
Request timed out"}
```

Can you understand the problem described in this event? The log record includes the `Request timed out` error message and has the `error` level, but it does not provide any meaningful context to us. Without any additional context, we can't easily understand the problem that caused the log event.

Providing the context of any log message is crucial for making it easy to work with the logs. This is especially important in a microservice environment, where similar operations can be performed by multiple services or components. It should always be easy to understand each message and have some reference to the component it is coming from. In this section, we already mentioned the practice of including some additional information, such as the name of the component, in a log event. Such metadata would generally include the following:

- Name of the service

- Name of the component emitting the event (for example, endpoint name)

- Name of the file (optional)

A more detailed version of the preceding log message looks like this:

```
{"level":"error", "time":"2022-09-04T20:10:10+1:00","message":"
Request timed out", "service":"rating", "component": "handler",
"endpoint": "putRating", "file": "handler.go"}
```

At this point, we have discussed the main topics related to logging and can move on to the next section, which describes another type of telemetry data – metrics.

Collecting service metrics

In this section, we are going to describe another type of service telemetry data: **metrics**. To understand what metrics are and how they are different from log data, let's start with an example. Imagine that you have a set of services providing APIs to their users, and you want to know how many times per second each API endpoint is called. How would you do this?

One possible way of solving this problem is using logs. We could create a log event for each request, and then we would be able to count the number of events for each endpoint, aggregating them by second, minute, or in any other possible way. Such a solution would work until we get too many requests per endpoint and can't log each one independently anymore. Let's assume there is a service that processes more than a million requests per second. If we used logs to measure its performance, we would need to produce more than a million log events every second, generating lots of data.

A more optimal solution to this problem would be to use some sort of value-based aggregation. Instead of storing the data representing each request separately, we could summarize the count of requests per second, minute, or hour, making the data more optimal for storing.

The problem that we just described is a perfect use case for using metrics – real-time quantitative measurements of system performance, such as request rate, latency, or cumulative counts. Like logs, metrics are time-based – each record includes a timestamp representing a unique instant of time in the past. However, unlike log events, metrics are primarily used for storing individual values. In our example, the value of an endpoint request rate metric would be the count of requests per second.

Metrics are generally represented as **time series** – sets of objects, called **data points**, containing the following data:

- Timestamp

- Value (most commonly, the value is numerical)

- An optional set of **tags**, defined as key-value pairs, that contain any additional metadata

To help you better understand the use cases of using metrics, let's define some common metric types:

- **Counters**: These are time series representing the value of a cumulative counter over time. An example would be the counter of service requests – each data point would include a timestamp and the count of requests at that particular moment.

- **Gauges**: These are time series representing the changes of a single scalar value over time. An example of a gauge is a dataset that contains the amount of free disk space on a server at different moments: each data point contains a single numerical value.

- **Histograms**: These are time series representing the distribution of some value against a predefined set of value ranges, called **buckets**. An example of a histogram metric is a dataset, containing the number of users for different age groups.

Let's focus on each metric type to help you understand their differences and the common use cases for each one.

Counter metrics are generally used for measuring two types of data:

- Cumulative value over time (for example, the total number of errors)

- Change of the cumulative value over time (for example, the number of newly registered users per hour)

The second use case is technically a different representation of the first one – if you know how many users you had at each moment in time, you can see how this value changes. Because of this, counters are often used to measure the rates of various events, such as API requests, over time.

The following code snippet provides an example of a `Counter` interface in a `tally` metrics library (we'll review this library later in this chapter):

```
type Counter interface {
    // Inc increments the counter by a delta.
    Inc(delta int64)
}
```

Unlike counters, gauges are used for storing unique values of measurements, such as the service's available memory over time. Here is a gauge example from a `tally` library:

```
type Gauge interface {
    // Update sets the gauges absolute value.
    Update(value float64)
}
```

Some of the other gauge use cases include the following:

- Number of goroutines running by a service instance
- Number of active connections
- Number of open files

Histograms are slightly different from counters and gauges. They require us to define a set of ranges that will be used to store the subsets of recorded data. The following are some examples of using histogram metrics:

- **Latency tracking**: You can track how long it takes to perform a certain service operation by creating a set of buckets representing various duration ranges. For example, your buckets could be 0–100 ms, 100–200 ms, 200–300 ms, and so on.
- **Cohort tracking**: You can track statistical data, such as the number of records in each group of values. For example, you can track how many users of each age subscribed to your service.

Now that we have covered some high-level basics of metrics, let's provide an overview of storing metrics.

Storing metrics

Similar to logs, storing metrics in a microservice environment brings some common challenges:

- **Collection and aggregation**: Metrics need to be collected from all service instances and sent for further aggregation and storage.
- **Aggregation**: Collected data needs to be aggregated, so various types of metrics, such as counters, would contain the data coming from all service instances. For example, the counter measuring the total number of requests should summarize the data across all service instances.

Let's review some of the popular tools that provide such features.

Prometheus

Prometheus is a popular open source monitoring solution that provides mechanisms for collecting and querying service metrics, as well as setting up automated alerts for detecting various types of incidents. Prometheus gained popularity in the developer community due to its simple data model and being a very flexible model for data ingestion, which we are going to cover in this section.

> **Note**
>
> Did you know that Prometheus is written in Go? You can check its source code on its GitHub page: `https://github.com/prometheus/prometheus`.

Prometheus supports three types of metrics – counters, gauges, and histograms. It stores each metric as a time series, similarly to the model that we described at the beginning of the *Collecting service metrics* section. Each metric contains the value, additional tags, called **labels**, and a name that can be used for identifying it.

Once the data gets into the Prometheus time series storage, it is available for querying via its query language, called **PromQL**. PromQL allows us to fetch time series data using various functions that allow us to easily filter or exclude certain name and label combinations. The following is an example of a PromQL query:

```
http_requests_total{environment="production",method!="GET"}
```

In this example, the query fetches time series with the `http_requests_total` name, a label that contains the environment key and production value, and any value of a method label that is not equal to `GET`.

There is an official Prometheus Go client on GitHub that provides various mechanisms to get the metrics data into Prometheus, as well as to execute the PromQL queries. You can access it here: `https://github.com/prometheus/client_golang`.

The documentation for instrumenting Go applications for using Prometheus can be found at the following link: `https://prometheus.io/docs/guides/go-application`.

Graphite

Graphite is another popular monitoring tool that offers metric collection, aggregation, and querying functionality that is similar to Prometheus. Although it has been among the oldest service monitoring tools in the industry, it remains an extremely powerful instrument for working with service metric data.

A typical Graphite installation consists of three main components:

- **Carbon**: A service that listens for time series data
- **Whisper**: A time series database
- **Graphite-web**: A web interface and an API for accessing the metrics data

Graphite offers a quick integration with a data visualization tool called **Grafana**, which we are going to cover in *Chapter 13* of this book. You can read more details about Graphite on its website: `https://graphiteapp.org`.

Now, let's move on to the next section, where we will describe the popular libraries for emitting service metrics.

Popular Go metrics libraries

There are some popular Go libraries for working with metrics that could help you ingest and query your time series metrics data. Let's provide a brief overview of some of them:

- **Tally** (`https://github.com/uber-go/tally`):

  - A performant and minimalistic library for emitting service metrics

  - Built-in support for data ingestion in Prometheus, StatsD, and M3

- **rcrowley/go-metrics** (`https://github.com/rcrowley/go-metrics`):

  - The Go port of a popular Java metric library (`https://github.com/dropwizard/metrics`)

  - Supports data ingestion into StatsD and Graphite

  - Has lots of integrations for exporting data to various observability systems, such as Datadog and Prometheus

- **go-kit/metrics** (`https://github.com/go-kit/kit/metrics`):

  - Part of the `go-kit` toolkit

  - Supports multiple metric storages, such as StatsD and Graphite

We will leave the decision of picking the metrics library for your services to you, as each library provides some useful features that you can leverage when developing your microservices. I am going to use the tally library in the examples throughout this chapter as it provides a simple and minimalistic API that can help illustrate the common metrics use cases. In the next section, we will review some of the use cases of using the metrics in the Go microservice code.

Emitting service metrics

In this section, we will provide some examples of emitting and collecting service metrics while covering some common scenarios, such as measuring API request rates, operation latencies, and emitting gauge values. We will use the tally library in our examples, but you can implement this logic using all other popular metric libraries.

First, let's provide an example of how to initialize the tally library so that you can start using it in your service code. In the following example, we are initializing it using the StatsD client (you can use any other tool to collect the metrics):

```
statter, err := statsd.
NewBufferedClient("127.0.0.1:8125","stats", time.Second, 1440)
if err != nil {
```

```
      panic(err)
}


reporter := tallystatsd.NewReporter(statter, tallystatsd.
Options{
      SampleRate: 1.0,
})
scope, closer := tally.NewRootScope(tally.ScopeOptions{
      Tags:      map[string]string{"service": "rating"},
      Reporter: reporter,
}, time.Second)
```

In this example, we are creating a tally reporter that will submit the metrics to the data collector (StatsD in our use case) and create a **scope** – an interface for reporting the metrics data – that would automatically submit them for collection.

Tally scopes are hierarchical: when we initialize the library, we create a **root scope**, which includes initial metadata in the form of key-value tags. All scopes that are created from it would include the parent metadata, preventing cases of missing tags during the metric's emission.

Once you get the scope, you can start reporting the metrics. The following example illustrates how to increment a counter metric by measuring the API request count, which would be automatically reported by tally:

```
counter := scope.Counter("request_count")
counter.Inc(1)
```

The `Inc` operation increments the value of the counter by 1, and the updated value of the metric gets collected by tally automatically in the background. This does not affect the performance of the function that performs the provided operations.

If you want to add some additional tags to the metric, you can use the `Tagged` function:

```
counter := scope.Tagged(map[string]string{"operation": "put"}).
Counter("request_count")
```

The following example illustrates how to update the gauge value. Let's say we have a function that calculates the number of active users in the system and we want to report this value to the metrics storage. We can achieve this by using a gauge metric in the following way:

```
gauge := scope.Gauge("active_user_count")
gauge.Update(userCount)
```

Now, let's provide an example of reporting a time duration. A common use case of this is reporting the latency of various operations. In the following example, we are reporting how long it takes to execute our function:

```
func latencyTrackingExample(scope tally.Scope) {
    timer := scope.Timer("operation_latency")
    stopwatch := timer.Start()
    defer stopwatch.Stop()
    // Function logic.
}
```

In our example, we are initializing the `operation_latency` timer and calling the `Start` function of it to start measuring operation latency. The `Start` function returns the instance of a `Stopwatch` interface, which includes the `Stop` function. This reports the time it takes since the stopwatch's start time.

When reporting the latency metrics, the tally library uses the default buckets, unless you provide their exact values. For example, when reporting the metrics to Prometheus, tally is using the following bucket configuration:

```
func DefaultHistogramBuckets() []float64 {
    return []float64{
        ms,
        2 * ms,
        5 * ms,
        10 * ms,
        20 * ms,
        50 * ms,
        100 * ms,
        200 * ms,
        500 * ms,
        1000 * ms,
        2000 * ms,
        5000 * ms,
        10000 * ms,
    }
}
```

Let's provide an example of using a histogram with a set of predefined numerical buckets:

```
histogram := scope.Histogram("user_age_distribution", tally.
MustMakeLinearValueBuckets(0, 1, 130))
histogram.RecordValue(userAgeInYears)
```

In our example, we are initializing the histogram metric using a set of predefined buckets from 0 to 130 and recording the value that matches the user age in years using it. Each bucket of the histogram will then contain the sum of the values.

Now that we have provided some basic examples of emitting service metrics, let's look at the best practices for working with metrics data.

Metrics best practices

In this section, we will describe some of the best practices related to metric data collection. The list is not exhaustive, but should still be useful for setting up metric collection logic in your services.

Keep tag cardinality in mind

When you emit your metrics and add additional tags to time series data, keep in mind that most time series databases are not designed to store **high-cardinality** data, which can contain lots of possible tag values. For example, the following types of data should not be generally included in service metrics:

- Object identifiers, such as movie or rating IDs
- Randomly generated data, such as UUIDs (for example, request UUIDs)

The reason for this is indexing is that each tag key-value combination must be indexed to make the time series searchable, and it becomes expensive to perform this when there are lots of distinct tag values.

You can still use some low-cardinality data in metric tags. The following are some possible examples:

- City ID
- Service name
- Endpoint name

Remember this tip to avoid reducing the throughput of your metrics pipeline and to ensure your services don't emit user identifiers and other types of high-cardinality metadata.

Standardize metric and tag names

Imagine that you have hundreds of services, and each service follows different metric naming conventions. For example, one service could be using `api_errors` as the name of the API error counter metric, while the other could be using the `api_request_errors` name. If you wanted to compare the metrics for such services, you would need to remember which naming convention each service was using. Such metric discovery will always take time, reducing your ability to analyze your data.

A much better solution is to standardize the names of common metrics and tags across all your services. This way, you can easily search and compare various performance indicators, such as service client and server error rates, API throughput, and request latency. In *Chapter 12*, we will review some common performance indicators that you can use to monitor the health of your services.

Set the appropriate retention

Most time series databases are capable of storing large datasets of metrics due to efficient aggregation. Unlike logs that require you to store each record independently, metrics can be aggregated into a smaller dataset. For example, if you store the counter data, you can store the sums of the values instead of storing each value separately. Even with these optimizations, time series data can take a lot of disk space to store. Large companies can store terabytes of metrics data, so it becomes important to manage its size and set up data retention policies, similar to logs and other types of telemetry data.

Metric storages, such as Prometheus, have a default retention time of 15 days, allowing you to change it in the settings. For example, to set the data retention time in Prometheus to 60 days, you can use the following flag:

```
--storage.tsdb.retention.time=60d
```

Limiting the storage retention time helps keep the size of the time series datasets under control, making it easier to manage the storage capacity and plan the infrastructure spending on data storage.

Now that we've discussed the metrics data, let's move on to the next section, which covers a powerful technique: tracing.

Tracing

So far, we have covered two common types of observability data – logs and metrics. Having logs and metrics data in place is often sufficient for service debugging and troubleshooting. However, there is another type of data that is useful for getting insights into microservice communication and data flows.

In this section, we are going to discuss **distributed tracing** – a technique that involves recording and analyzing interactions between different services and service components. The main idea behind distributed tracing is to automatically record all such interactions and provide a convenient way to visualize them. Let's look at the following example, which illustrates a distributed tracing use case known as call analysis:

Figure 11.2 – Tracing visualization example

Here, you can see the execution of a single `GetMovieDetails` request for our movie service. The data provides some insights into the operation's execution:

- Soon after the request starts, two parallel calls come from the movie service; one to the metadata service and the other to the rating service.

- The call to the metadata service takes 100 milliseconds to complete.

- The call to the rating service takes 1,100 milliseconds to complete, spanning almost the entire request processing time.

The data that we just extracted provided us with lots of valuable information for analyzing the movie service's performance. First, it helped us understand how an individual request was handled by the movie service, as well as which sub-operations it performed. We can also see the duration of each operation and find out which one was slowing down the entire request. By using this data, we could troubleshoot endpoint performance issues, finding which components make a significant impact on request processing.

In our example, we showed the interaction between just three services, but tracing tools allow us to analyze the behavior of systems that use tens and even hundreds of services simultaneously. The following are some other use cases that make tracing a powerful tool for production debugging:

- **Error analysis**: Tracing allows us to visualize the errors on complex call paths, such as chains of calls spanning lots of different services.

- **Call path analysis**: Sometimes, you may investigate issues in systems you are not very familiar with. Tracing data helps you visualize the call path of various operations, helping you understand the logic of the services without requiring you to analyze their code.

- **Operation performance breakdown**: Tracing allows us to see the duration of individual steps of a long-running operation.

Let's describe the tracing data model so that you can get familiar with its common terminology. The core element of tracing is a **span** – a record representing some logical operation, such as a service endpoint call. Each span has the following properties:

- Operation name

- Start time

- End time

- An optional set of tags providing some additional metadata related to the execution of the associated operation

- An optional set of associated logs

Spans can be grouped into hierarchies to model relationships between different operations. For example, in *Figure 11.2*, our GetMovieDetails span included two child spans, representing GetMetadata and GetAggregatedRating operations.

Let's explore how we can collect and use tracing data in our Go applications.

Tracing tools

There are various tools for distributed tracing in a microservice environment. Among the most popular ones is Jaeger, which we will review in this section.

> **Note**
>
> There are multiple observability libraries and tools available for developers, including libraries for trace collection. To standardize the data model and make data interchangeable, there is an OpenTelemetry project that contains the specification of the tracing data model. You can get familiar with the project on its website: `https://opentelemetry.io`.

Jaeger is an open source distributed tracing tool that provides mechanisms for collecting, aggregating, and visualizing the tracing data. It offers a simple but highly flexible setup and a great user interface for accessing trace data. This is the reason why it quickly became one of the most popular observability tools across the industry.

Jaeger is compatible with the OpenTelemetry specification, so it can be used in combination with any clients implementing the tracing specification, such as the Go SDK (`https://opentelemetry.io/docs/instrumentation/go/`). The OpenTelemetry SDK is currently the recommended way of emitting trace data from applications, so we are going to use it in our examples throughout this section.

The general data flow for services using Jaeger looks like this:

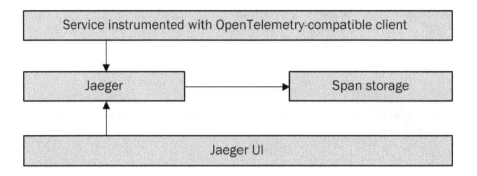

Figure 11.3 – Jaeger data flow

In the preceding diagram, a service using a Jaeger-compatible library is emitting traces to the Jaeger backend, which is storing them in the span storage. The data is accessible for querying and visualization via the Jaeger UI.

> **Note**
>
> Jaeger is another example of an observability tool written in Go. You can check out the Jaeger source code on its official GitHub page: `https://github.com/jaegertracing/jaeger`.

You can find more information about the Jaeger project on its website: `https://www.jaegertracing.io`.

Let's see some examples of instrumenting Go services to emit the tracing data. We will use the OpenTelemetry SDK in our examples to make our code compatible with different tracing software.

Collecting tracing data with the OpenTelemetry SDK

In this section, we will show you how to emit tracing data in your service code.

As we mentioned at the beginning of the *Tracing* section, the core benefit of distributed tracing is the ability to automatically capture data that shows how services and other network components communicate with each other. Unlike metrics, which measure the performance of individual operations, traces help to collect information on how each request or operation is handled across the entire network of nodes that report this data. To report the traces, service instances need to be **instrumented** so that they perform two distinct roles:

- **Report the data on distributed operations**: For each **traceable operation** – an operation spanning multiple components, such as network requests or database queries – an instrumented service should report spans. The report should contain the operation name, start time, and end time.

- **Facilitate context propagation**: The service should explicitly propagate context throughout the execution (if you are confused by this, please read the following paragraphs – this is the main trick behind distributed tracing!).

We already defined a span earlier in this chapter, so let's move to the second requirement for service instrumentation. What is context propagation and how do we perform it in the Go microservice code?

Context propagation is a technique that involves explicitly passing an object, called a **context**, into other functions in the form of an argument. The context may contain arbitrary metadata, so passing it to another function helps propagate it further. That is, each function down the stream can either add new metadata to the context or access the metadata that already exists in it.

Let's illustrate context propagation via a diagram:

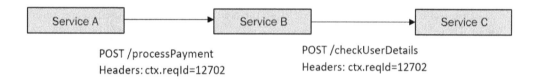

Figure 11.4 – Context propagation example

In the previous flow chart, there is an HTTP request coming from **Service A** to **Service B** to process a payment. **Service A** includes an additional HTTP header called `ctx.reqId` for the request to pass the request identifier. **Service B** calls **Service C** to check the user details to verify whether the user is eligible for making a payment. Then, **Service B** passes the `ctx.reqId` header further to **Service C** so that all services can record the identifier of the request, for which they perform the operations.

The example that we just provided illustrates context propagation between three services. This is achieved by including specific HTTP headers in requests, which provide additional metadata for request processing. There are multiple ways of propagating the data when executing various operations. We will start by looking at the regular Go function calls.

We covered Go context propagation in *Chapter 2* and mentioned the `context` package, which provides a type called `context.Context`. Passing the context between two Go functions of a single service is as easy as calling another function with an additional argument, as shown here:

```
func ProcessRequest(ctx context.Context, ...) {
    return ProcessAnotherRequest(ctx, ...)
}
```

In our example, we pass the context that we receive in our function into another one, propagating it throughout the execution chain. We can attach additional metadata to the context by using the `WithValue` function, as shown in the following code block:

```
func ProcessRequest(ctx context.Context, ...) {
    newCtx = context.WithValue(ctx, someKey, someValue)
    return ProcessAnotherRequest(newCtx, ...)
}
```

In this updated example, we are passing the modified context to the other function, which will include some additional tracing metadata.

Now, let's connect this knowledge with the core concept of tracing – a span. A span represents an individual operation, such as a network request, that can be related to some other operations, such as the other network calls that are made during the request's execution. In our `getMovieDetails` example, the original request would be represented as a **root span** or a parent span. Its child spans represent the calls to the endpoints of metadata and rating services – both calls are made as a part of `getMovieDetails` request handling. To establish the relationship between the child and parent spans, we need to pass the identifier of the parent span to its children. We can do this by propagating it through the context of each function call, as we illustrated earlier. To make this easier to understand, let's summarize the steps for collecting the trace data for a Go function:

1. For the original function being traced, we generate a new parent span object.
2. When the function makes calls to any other functions that need to be included in the trace (for example, network calls or database requests), we pass the parent span data to them as a part of the Go `context` argument.
3. When a function receives a context with some parent span metadata, we include the parent span ID in the span data associated with the function.
4. All the functions in the chain should follow the same steps and, at the end of each execution, report the captured span data.

Now, let's demonstrate how to use this technique in Go applications. We are going to use the OpenTelemetry Go SDK in our examples, and use Jaeger as the data source of the tracing data:

1. Let's start with the configuration changes. Inside each service directory, update the `cmd/config.go` file to the following:

```
package main

type config struct {
    API    apiConfig    `yaml:"api"`
    Jaeger jaegerConfig `yaml:"jaeger"`
```

```
}

type apiConfig struct {
    Port int `yaml:"port"`
}


type jaegerConfig struct {
    URL string `yaml:"url"`
}
```

The configuration that we just added will help us set the Jaeger URL for submitting the trace data.

2. The next step is to update the `configs/base.yaml` file for each service so that it includes the Jaeger API URL property. We can do this by adding the following code at the end:

```
jaeger:
  url: http://localhost:14268/api/traces
```

3. Let's create a shared function that can be used in each service to initialize the tracing data provider. This is going to submit our traces to Jaeger. In our root pkg directory, create a directory called `tracing` and add a `tracing.go` file with the following contents:

```
package tracing

import (
    "go.opentelemetry.io/otel/exporters/jaeger"
    "go.opentelemetry.io/otel/sdk/resource"
    tracesdk "go.opentelemetry.io/otel/sdk/trace"
    semconv "go.opentelemetry.io/otel/semconv/v1.12.0"
)


// NewJaegerProvider returns a new jaeger-based tracing
provider.
func NewJaegerProvider(url string, serviceName string)
(*tracesdk.TracerProvider, error) {
    exp, err := jaeger.New(jaeger.
WithCollectorEndpoint(jaeger.WithEndpoint(url)))
    if err != nil {
        return nil, err
    }
```

```
    tp := tracesdk.NewTracerProvider(
        tracesdk.WithBatcher(exp),
        tracesdk.WithResource(resource.NewWithAttributes(
            semconv.SchemaURL,
            semconv.ServiceNameKey.String(serviceName),
        )),
    )
    return tp, nil
}
```

Here, we are initializing the Jaeger client and using it to create the OpenTelemetry trace data provider. The provider will automatically submit the trace data that we will collect throughout the service execution.

4. The next step is to update the `main.go` file of each service. Add the `go.opentelemetry.io/otel` import to the `imports` block of the `main.go` file for each service, and add the following code block after the first `log.Printf` call:

```
    tp, err := tracing.NewJaegerProvider(cfg.Jaeger.URL,
serviceName)
    if err != nil {
        log.Fatal(err)
    }
    defer func() {
        if err := tp.Shutdown(ctx); err != nil {
            log.Fatal(err)
        }
    }()
    otel.SetTracerProvider(tp)
    otel.SetTextMapPropagator(propagation.TraceContext{})
```

The last two lines of the code set the global OpenTelemetry trace provider to our Jaeger-based version. These lines also enable context propagation, which will allow us to transfer the tracing data between the services.

5. To enable client-side context propagation, update the `internal/grpcutil/grpcutil.go` file to the following:

```
package grpcutil

import (
```

```
    "context"
    "math/rand"

    "go.opentelemetry.io/contrib/instrumentation/google.
golang.org/grpc/otelgrpc"
    "google.golang.org/grpc"
    "google.golang.org/grpc/credentials/insecure"
    "movieexample.com/pkg/discovery"
)


// ServiceConnection attempts to select a random service
// instance and returns a gRPC connection to it.
func ServiceConnection(ctx context.Context, serviceName
string, registry discovery.Registry) (*grpc.ClientConn,
error) {
    addrs, err := registry.ServiceAddresses(ctx,
serviceName)
    if err != nil {
        return nil, err
    }
    return grpc.Dial(
        addrs[rand.Intn(len(addrs))],
        grpc.WithTransportCredentials(insecure.
NewCredentials()),
        grpc.WithUnaryInterceptor(otelgrpc.
UnaryClientInterceptor()),
    )
}
```

Here, we added an OpenTelemetry-based interceptor that injects the tracing data into each request.

6. Inside the main.go file of each service, change the line containing the grpc.NewServer() call to the following one, to enable server-side context propagation:

```
srv := grpc.NewServer(grpc.UnaryInterceptor(otelgrpc.
UnaryServerInterceptor()))
```

The change that we just made is similar to the previous step, just for server-side handling.

7. The last step is to make sure all the new libraries are included in our project by running the following command:

```
go mod tidy
```

With that, our services have been instrumented with tracing code and emit span data on each API request.

Let's test our newly added code by running our services and making some requests to them:

1. To be able to collect the tracing data, you will need to run Jaeger locally. You can do this by running the following command:

```
docker run -d --name jaeger \
  -e COLLECTOR_OTLP_ENABLED=true \
  -p 6831:6831/udp \
  -p 6832:6832/udp \
  -p 5778:5778 \
  -p 16686:16686 \
  -p 4317:4317 \
  -p 4318:4318 \
  -p 14250:14250 \
  -p 14268:14268 \
  -p 14269:14269 \
  -p 9411:9411 \
  jaegertracing/all-in-one:1.37
```

2. Now, we can start all our services locally by executing the go run *.go command inside each cmd directory.

3. Let's make some requests to our movie service. In *Chapter 5*, we mentioned the grpcurl tool. Let's use it again to make a manual gRPC query:

```
grpcurl -plaintext -d '{"movie_id":"1"}' localhost:8083
MovieService/GetMovieDetails
```

If everything was correct, we should get our trace in Jaeger. Let's check it in the Jaeger UI by going to http://localhost:16686/. You should see a similar page as follows:

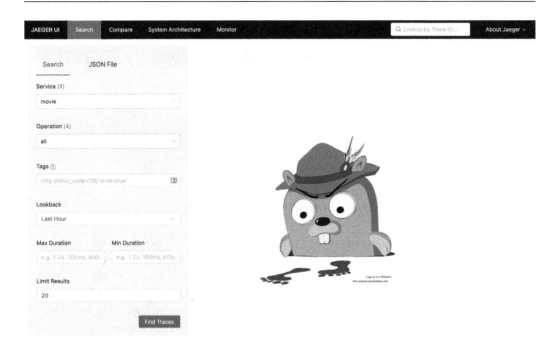

Figure 11.5 – Jaeger UI

4.   Select the **movie** service in the **Service** field and click **Find Traces**. You should see some trace results, as shown here:

Figure 11.6 – Jaeger traces for the movie service

5.   If you click on the trace, you will see its visualized view, as shown here:

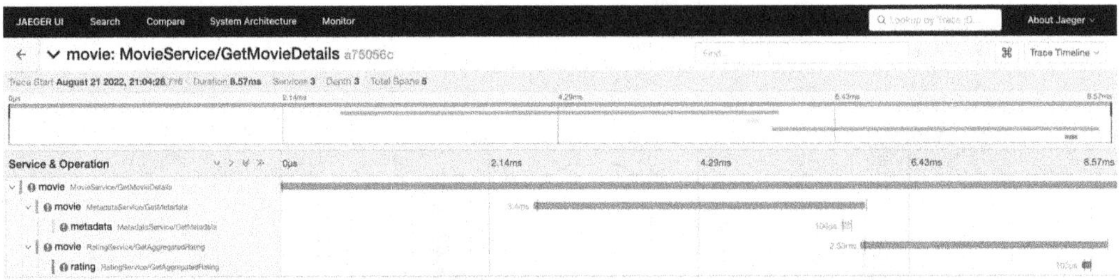

Figure 11.7 – Jaeger trace view for the GetMovieDetails endpoint call

On the left panel, you can see the request as a tree of spans, where the root span represents the MovieService/GetMovieDetails operation, which includes the calls to the MetadataService/GetMetadata and RatingService/GetAggregatedRating endpoints. Congratulations, you have set up distributed tracing for your microservices using the OpenTracing SDK! All our gRPC calls are now traced automatically, without any need to add any extra service logic. This provides us with a convenient mechanism for collecting valuable data on service communication.

As an extra step, let's illustrate how to add tracing for our database operations. As you can see from the trace view in the preceding screenshot, we currently don't have any database-related spans on our graph. This is because our database logic has not been instrumented yet. Let's demonstrate how to do this manually:

1.   Open the metadata/internal/repository/memory/memory.go file and add go.opentelemetry.io/otel to its imports.

2.   In the same file, add the following constant:

```
const tracerID = "metadata-repository-memory"
```

3.   At the beginning of the Get function, add the following code:

```
_, span := otel.Tracer(tracerID).Start(ctx, "Repository/
Get")
defer span.End()
```

4.   Add a similar code block at the beginning of the Put function:

```
_, span := otel.Tracer(tracerID).Start(ctx, "Repository/
Put")
defer span.End()
```

We just manually instrumented our in-memory metadata repository for emitting the trace data on its primary operations, `Get` and `Put`. Now, each call to these functions should create a span in the captured trace, allowing us to see when and how long each operation is being executed.

Let's test our newly added code. Restart the metadata service and make a new `grpcurl` request to the movie service provided previously. If you check for new traces in Jaeger, you should see the new one, with an additional span:

Figure 11.8 – Jaeger trace view with an additional repository span

Notice the last span in the trace view, representing the `Repository/Get` operation. It is the result of our change. Now, we can see the database operations on our traces. You can go ahead and update the rating service repository by including similar logic – follow the preceding instructions, and you should be able to make it work in the same way that we just did for the metadata service.

When should you manually add span data to your functions? I would suggest doing this for each operation involving network calls, I/O operations (such as writing and reading from files), database writes and reads, and any other calls that can take a substantial amount of time. I would personally say that any function that takes more than 50 ms to complete is a good candidate for tracing.

At this point, we have provided a high-level overview of Go tracing techniques, and this marks an end to our journey into telemetry data. In the next few chapters, we will continue our explorations into other fields, such as dashboarding, system-level performance analysis, and some advanced observability techniques.

Summary

In this chapter, we covered observability by describing various techniques for analyzing the real-time performance of Go microservices and covering the main types of service telemetry data, such as logs, metrics, and traces. You learned about some of the best practices for performing logging, metric collection, and distributed tracing. We demonstrated how you can instrument your Go services to collect the telemetry data, as well as how to set up the tooling for distributed tracing. We also provided some examples of tracing the requests spanning three of the services that we implemented earlier in this book.

The knowledge that you gained in this chapter should help you debug various performance issues of your microservices, as well as enable monitoring of various types of telemetry data. In *Chapter 12*, we will demonstrate how to use the collected telemetry data to set up service alerting for detecting service-related incidents as quickly as possible.

Further reading

To learn more about the topics that were covered in this chapter, take a look at the following resources:

- *Monitoring Distributed Systems*: `https://sre.google/sre-book/monitoring-distributed-systems/`

- *Effective Troubleshooting*: `https://sre.google/sre-book/effective-troubleshooting/`

- *Mastering Distributed Tracing*: `https://www.packtpub.com/product/mastering-distributed-tracing/9781788628464`

- Logging best practices: `https://devcenter.heroku.com/articles/writing-best-practices-for-application-logs`

- *Ten commandments of logging*: `https://www.dataset.com/blog/the-10-commandments-of-logging/`

- Microservice logging tips: `https://www.techtarget.com/searchapparchitecture/tip/5-essential-tips-for-logging-microservices`

- OpenTelemetry documentation: `https://opentelemetry.io/docs/`

- Beginner's Guide to OpenTelemetry: `https://logz.io/learn/opentelemetry-guide/`

- *The 3 Pillars of System Observability*: `https://iamondemand.com/blog/the-3-pillars-of-system-observability-logs-metrics-and-tracing/`

- *What is observability?*: `https://www.dynatrace.com/news/blog/what-is-observability-2/`

- *What is Telemetry?*: `https://www.sumologic.com/insight/what-is-telemetry/`

12

Setting Up Service Alerting

In the previous chapter, we described various types of service telemetry data, such as logs, metrics and traces, and illustrated how to collect them for troubleshooting service performance issues.

In this chapter, we will illustrate how to use telemetry data to automatically detect incidents by setting up alerts for our microservices. You will learn which types of service metrics to collect, how to define the conditions for various incidents, and how to establish the complete alerting pipeline for your microservices using a popular monitoring and alerting tool, Prometheus.

We will cover the following topics:

- Alerting basics

- Introduction to Prometheus

- Setting up Prometheus alerting for our microservices

- Alerting best practices

Now, we are going to proceed to the overview of alerting basics.

Technical requirements

To complete this chapter, you will need Go 1.11+ or above. You will also need the Docker tool, which you can download at `https://www.docker.com/`.

You can find the code examples for this chapter on GitHub: `https://github.com/PacktPublishing/microservices-with-go/tree/main/Chapter12`.

Alerting basics

No microservice operates without incidents; even if you have a stable, highly tested, and well-maintained service, it can still experience various types of issues, such as the following:

- **Resource constraints**: A host running the service may experience high CPU utilization or insufficient RAM or disk space.

- **Network congestion**: The service may experience a sudden increase in load or decreased performance in any of its dependencies. This could limit its ability to process incoming requests or operate at the expected performance level.

- **Dependency failures**: Other services or libraries that your service is depending on may experience various issues, affecting your service execution.

Such issues can be self-resolving. For example, a slower network throughput could be a transient issue caused by temporary maintenance or a network device being restarted. Many other types of issues, which we call incidents, require some actions from the engineers to be mitigated.

To mitigate an incident, first, we need to detect it. Once the issue is known, we can notify the engineers or perform automated actions, such as an automated deployment rollback or application restart. In this chapter, we will describe the **alerting technique** that combines incident detection and notification. This technique can be used to automate the incident response to various types of microservice issues.

The key principles behind alerting are pretty simple and can be summarized by the following statements:

- To set up alerts, developers define the **alerting conditions.**

- Alerting conditions are based on the telemetry data (most commonly, metrics) and are defined in the form of queries.

- Each defined alerting condition is evaluated periodically, such as every minute.

- If the alerting condition is met, the associated actions are executed (for example, an email or an SMS is sent to an engineer).

To illustrate how alerting works, imagine that one of your services is emitting a metric called `active_user_count` that reports the number of active users at a particular moment. Let's assume that we would like to get notified if the number of active users suddenly drops to zero. Such a situation would likely indicate some incident with our service unless we have too few users (for simplicity, we will assume our system should always have some active users).

Using pseudocode, we could define the alerting condition for our use case in the following way:

```
active_user_count == 0
```

Once the alerting condition has been met, the alerting software would check actions that should be triggered based on its configuration. Assuming that we have configured our alerts to trigger email notifications, it would send the emails and include any necessary metadata. The metadata would include information such as which incident just occurred and, if provided, the steps to mitigate it.

We will provide some examples of alerting configurations later in this chapter. For now, we will focus on some practical use cases, providing you with some ideas for setting up alerting for your services.

Alerting use cases

There are many use cases for which you would need to set up automated alerts. In this section, we will provide some common examples that can act as a reference point for you.

In the *Google SRE* book we mentioned earlier in *Chapter 10*, there was a definition of **The Four Golden Signals** of monitoring, which can be used to monitor various types of applications, from microservices to data processing pipelines. These signals provide a great basis for service alerting, so let's review them and describe how you can use each one to increase your service reliability:

- **Latency**: Latency is a measure of processing time, such as the duration of processing an API request, a Kafka message, or any other operation. It is the main indicator of system performance – when it gets too high, the system starts affecting its callers, creating network congestion. You should generally track the latency of your primary operations, such as API endpoints providing the critical functionality.

- **Traffic**: Traffic measures the load on your system, such as the number of requests your microservices are getting at the current moment. An example of a traffic-based metric is an API request rate, measured as the number of requests per second. Measuring traffic is important for ensuring you have enough capacity to handle the requests to your system.

- **Errors**: Errors are often measured as the **error rate** or the ratio between the failed and total operations. Measuring the error rate is critical for ensuring your services remain operational.

- **Saturation**: Saturation generally measures the utilization of your resources, such as RAM or disk usage, CPU, or I/O load. You should keep track of saturation to ensure your services don't fail unexpectedly due to resource insufficiency.

These Four Golden Signals can help you establish monitoring and alerting for your services and critical operations, such as your primary API endpoints. Let's provide some practical examples to help you understand some common alerting use cases.

First, let's start with the common signals for API alerting that can be measured either across all endpoints or on a per-endpoint basis:

- **API client error rate**: The ratio between the requests that fail due to client errors and all requests

- **API server error rate**: The ratio between the requests that fail due to server errors and all requests

- **API latency**: The time it takes to process requests

Now, let's provide some examples of signals for measuring system saturation:

- **CPU utilization**: How much your CPUs are being used on a scale from 0% (unused/idle) to 100% (fully used, no extra capacity).

- **Memory utilization**: Ratio between the used and total memory.

- **Disk utilization**: Percentage of used disk space.

- **Open file descriptors**: File descriptors are often used to handle network requests, file writes and reads, and other I/O operations. There is usually a limit on the number of open file descriptors per process, so if your service reaches a critical limit (based on your OS settings), your service may fail to serve requests.

Let's also provide some examples of other signals to monitor:

- **Service panics**: The general recommendation is not to tolerate any service panics, as they often signal application bugs or issues such as out-of-memory errors.

- **Failed deployments**: You can automate the detection of failed deployments and emit a metric indicating the failure, using it to create automated alerts.

Now that we have covered some common alerting use cases, let's proceed to the overview of Prometheus, which we will use to set up our microservice alerts.

Introduction to Prometheus

In *Chapter 11*, we mentioned a popular open source alerting and monitoring tool called Prometheus that can collect service metrics and set up automated alerts based on the metric data. In this section, we will demonstrate how to use Prometheus to set up alerts for our microservices.

Let's summarize our learning about Prometheus from *Chapter 11*:

- Prometheus allows us to collect and store service metrics in the form of a time series.

- There are three types of metrics – counters, histograms, and gauges.

- To query metrics data, Prometheus offers a query language called PromQL.

- Service alerts can be configured using a tool called Alertmanager.

Metrics can be imported from service instances into Prometheus in two different ways:

- **Scraping**: Prometheus reads metrics from service instances.

- **Pushing**: The service instance sends metrics to Prometheus using a dedicated service, the Prometheus Pushgateway.

Scraping is the recommended way of setting up metric data ingestion in Prometheus. Each service instance needs to expose an endpoint to provide the metrics, and Prometheus takes care of pulling the data and storing it for further querying, as shown in the following diagram:

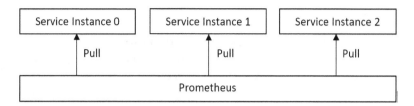

Figure 12.1 – Prometheus scraping model

Let's provide an example of a service instance response to a scraping request by Prometheus. Let's assume you add a separate HTTP API endpoint called /metrics and return the newest service instance metrics in the following format:

```
active_user_count 755
api_requests_total_count 18900
api_requests_getuser_count 500
```

In this example, the service instance reports three metrics in the form of key-value pairs, where the key defines a time series name and the value defines the value of the time series at the current moment. Once Prometheus calls the /metrics endpoint, the service instance should provide a new dataset containing only time series that have not been included in previous responses.

Once Prometheus collects the metrics, they become available for querying using a Prometheus-specific language called PromQL. PromQL-based queries can be used to analyze the time series data through the Prometheus UI or to set up automated alerts using Alertmanager. For example, the following query returns all values of the active_user_count time series, as well as their tags:

```
active_user_count
```

You can use additional query filters, called **matchers**, to include only specific data points. For example, if multiple services emit the active_user_count metric, you can only request time series that have a particular tag value:

```
active_user_count{service="rating-ui"}
```

Alerting conditions are generally defined as expressions that return Boolean results. For example, to define the alerting condition when the active user count drops to zero, you would use the following PromQL query with the == operator:

```
active_user_count == 0
```

The PromQL language provides some other types of time series matchers, such as quantile, which can be used to perform various aggregations. The following query example can be used to check whether the median api_request_latency value exceeds 1:

```
api_request_latency{quantile="0.5"} > 1
```

You can become familiar with the other aspects of the PromQL language by reading the official documentation on its website: https://prometheus.io/docs/prometheus/latest/querying/basics/. Now, let's explore how to set up alerts using the Prometheus alerting tool, Alertmanager.

Alertmanager is a separate component of Prometheus that allows us to configure alerts and notifications to detect various types of incidents. Alertmanager operates by reading the provided configuration and querying Prometheus time series data periodically. Let's provide an example of Alertmanager's configuration:

```
groups:
- name: Availability alerts
  rules:
  - alert: Rating service down
    expr: service_availability{service="rating"} == 0
    for: 3m
    labels:
      severity: page
    annotations:
      title: Rating service availability down
      description: No available instance of the rating service.
```

In our configuration example, we set an alert for when the value of the service_availability metric, which has a service="rating" tag, is equal to 0 for 3 minutes or more, triggering a PagerDuty incident to notify the on-call engineer about the issue.

Some other features of Alertmanager include notification grouping, notification retries, and alert suppression. To illustrate how Prometheus and Alertmanager work in practice, let's describe how to set them up for our example microservices from the previous chapters.

Setting up Prometheus alerting for our microservices

In this section, we will illustrate how to set up service alerting using Prometheus and its alerting extension, Alertmanager, for the services we created in previous chapters. You will learn how to expose the service metrics for collection, how to set up Prometheus and Alertmanager to aggregate and store the metrics from multiple services, and how to define and process service alerts.

Our high-level approach is as follows:

1. Set up Prometheus metric reporting to our services.

2. Install Prometheus and configure it to scrape the data from the three example services that we created in previous chapters.

3. Configure service availability alerts using Alertmanager.

4. Test our alerts by triggering an alerting condition and running Alertmanager.

Let's start by illustrating how to integrate our services with Prometheus. To do this, we need to add a metric collection to our services by exposing an endpoint that will provide the newest metrics to Prometheus.

First, we need to add Prometheus configuration to our services. In each service directory, update the cmd/config.go file to the following:

```go
package main

type config struct {
    API        apiConfig        `yaml:"api"`
    Jaeger     jaegerConfig     `yaml:"jaeger"`
    Prometheus prometheusConfig `yaml:"prometheus"`
}

type apiConfig struct {
    Port int `yaml:"port"`
}

type jaegerConfig struct {
    URL string `yaml:"url"`
}

type prometheusConfig struct {
```

```
        MetricsPort int `yaml:"metricsPort"`
}
```

Our new configuration allows us to specify the service port of the metric collection endpoint. Inside each `configs/base.yaml` file, add the following block:

```
prometheus:
    metricsPort: 8091
```

We are ready to update our services so that they can start reporting the metrics. Update the main. go file of each service by adding the following imports:

```
    "github.com/uber-go/tally"
    "github.com/uber-go/tally/prometheus"
```

In any part of the main function, add the following code:

```
    reporter := prometheus.NewReporter(prometheus.Options{})
    _, closer := tally.NewRootScope(tally.ScopeOptions{
        Tags:               map[string]string{"service":
"metadata"},
        CachedReporter: reporter,
    }, 10*time.Second)
    defer closer.Close()
    http.Handle("/metrics", reporter.HTTPHandler())
    go func() {
        if err := http.ListenAndServe(fmt.Sprintf(":%d", cfg.
Prometheus.MetricsPort), nil); err != nil {
            logger.Fatal("Failed to start the metrics handler",
zap.Error(err))
        }
    }()

    counter := scope.Tagged(map[string]string{
        "service": "metadata",
    }).Counter("service_started")
    counter.Inc(1)
```

In the code we just added, we initialized the `tally` library to collect and report the metrics data, which we mentioned in *Chapter 11* of this book. We used a built-in Prometheus reporter that implements metric data collection using the Prometheus time series format and exposed an HTTP endpoint to allow Prometheus to collect our data.

Let's test the newly added endpoint. Restart the metadata service and try accessing the new endpoint by opening `http://localhost:8091/metrics` in your browser. You should get a similar response:

```
# HELP go_gc_duration_seconds A summary of the pause duration
of garbage collection cycles.
# TYPE go_gc_duration_seconds summary
go_gc_duration_seconds{quantile="0"} 0
go_gc_duration_seconds{quantile="0.25"} 0
...
```

The response of the metrics handler includes the Go runtime data, such as the number of goroutines at the current moment, the Go library version, and many other useful metrics.

Now, we are ready to set up Prometheus alerting. Inside the `src` directory of our project, create a directory called `configs` and add a `prometheus.yaml` file with the following contents:

```
global:
  scrape_interval: 15s
  scrape_timeout: 10s
  evaluation_interval: 15s
alerting:
  alertmanagers:
  - follow_redirects: true
    enable_http2: true
    scheme: http
    timeout: 10s
    api_version: v2
    static_configs:
    - targets:
      - host.docker.internal:9093
```

Additionally, add the following configuration to the file:

```
rule_files:
- alerts.rules
scrape_configs:
- job_name: prometheus
  honor_timestamps: true
  scrape_interval: 15s
  scrape_timeout: 10s
  metrics_path: /metrics
  scheme: http
  follow_redirects: true
  enable_http2: true
  static_configs:
  - targets:
    - localhost:9090
  - targets:
    - host.docker.internal:8091
    labels:
      service: metadata
  - targets:
    - host.docker.internal:8092
    labels:
      service: rating
  - targets:
    - host.docker.internal:8093
    labels:
      service: movie
```

Let's describe the configuration that we just added. We set the scraping interval provided to 15 seconds and provided a set of targets to scrape the metrics data, which includes the address of each of our services. You may notice that we are using the host.docker.internal network address in each target definition — we will run Prometheus using Docker and the host.docker.internal address will allow it to access our newly added endpoints running outside of Docker.

Note that we provided a static list of service addresses inside the static_configs block. We did this intentionally to illustrate the simplest scraping approach, which is when Prometheus knows the address of each service instance. In a dynamic environment, where service instances can be added or removed, you would need to use Prometheus with a service registry, such as Consul. Prometheus provides built-in support for scraping metrics from services registered with Consul: instead of `static_configs`, you could define the Consul scraping configuration:

```
consul_sd_configs:
- server:  host.docker.internal:8500
  services:
    - <SERVICE_NAME>
```

Next, we will demonstrate how to scrape a static list of service instances; you can try setting up Consul-based Prometheus scraping as an additional exercise after reading this chapter. Let's add alerting rules for our services. Inside the newly added `configs` directory, create the `alerts.rules` file and add the following to it:

```
groups:
- name: Service availability
  rules:
  - alert: Metadata service down
    expr: up{service="metadata"} == 0
    labels:
      severity: warning
    annotations:
      title: Metadata service is down
      description: Failed to scrape {{ $labels.service }}.
Service possibly down.
  - alert: Rating service down
    expr: up{service="rating"} == 0
    labels:
      severity: warning
    annotations:
      title: Metadata service is down
      description: Failed to scrape {{ $labels.service }}
service on {{ $labels.instance }}. Service possibly down.
  - alert: Movie service down
    expr: up{service="movie"} == 0
    labels:
      severity: warning
    annotations:
      title: Metadata service is down
```

```
        description: Failed to scrape {{ $labels.service }}
  service on {{ $labels.instance }}. Service possibly down.
```

The file we just added includes the alert definitions for each of our services. Each alert definition includes the expression Prometheus would check to evaluate whether an associated alert should be fired.

Now, we are ready to install and run Prometheus to test our alerting. Inside the `src` directory of our project, run the following command to run Prometheus using the newly created configuration:

```
docker run \
    -p 9090:9090 \
    -v configs:/etc/prometheus \
    prom/prometheus
```

If everything is successful, you should be able to access the Prometheus UI by opening `http://localhost:9090/`. On the initial screen, you will see the search input you can use to access the Prometheus metrics emitted by our services. Type `up` into the search input and click **Execute** to access the metrics:

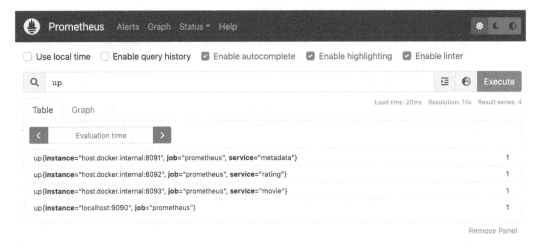

Figure 12.2 – Prometheus metrics search

You can go to the **Alerts** tab to see the currently configured alerts that we defined in our `alerts.rules` file:

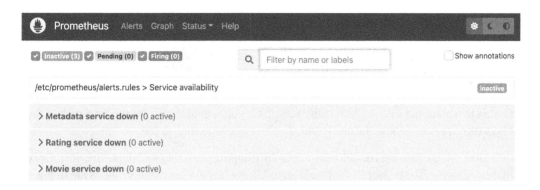

Figure 12.3 – Prometheus Alerts view

If all three services are running, all three associated alerts should be marked as **inactive**. We will get back to the **Alerts** page shortly; for now, let's proceed and set up Alertmanager so that we can trigger some alerts for our services.

Inside our `configs` directory, including the Prometheus configuration, add a file called `alertmanager.yml` with the following contents:

```
global:
  resolve_timeout: 5m
route:
  repeat_interval: 1m
  receiver: 'email'
receivers:
- name: 'email'
  email_configs:
  - to: 'your_email@gmail.com'
    from: 'your_email@gmail.com'
    smarthost: smtp.gmail.com:587
    auth_username: 'your_email@gmail.com'
    auth_identity: 'your_email@gmail.com'
    auth_password: 'your_password'
```

Update the email configuration in the file we just created so that Alertmanager can send some emails for our alerts.

Now, run the following command to start Alertmanager:

```
docker run -p 9093:9093 -v <PATH_TO_CONFIGS_DIR>:/etc/
alertmanager prom/alertmanager --config.file=/etc/alertmanager/
alertmanager.yml
```

Don't forget to replace the <PATH_TO_CONFIGS_DIR> placeholder with the full local path to the configs directory containing the newly added alertmanager.yml file.

Now, let's simulate the alerting condition by manually stopping the rating and movie services. Once you do this, open the **Alerts** page in the Prometheus UI; you should see that both alerts are **firing**:

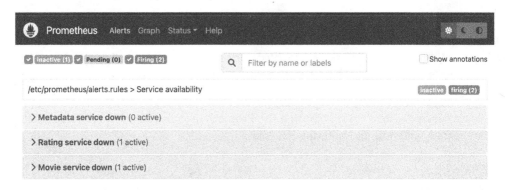

Figure 12.4 – Firing Prometheus alerts

You can access the Alertmanager UI by going to http://localhost:9093.

If alerts are fired in Prometheus, you should also see them in the Alertmanager UI:

Figure 12.5 – The Alertmanager UI

If you configured Alertmanager correctly, you should get an email to the address you provided in the configuration. If you haven't received an email, check the Docker logs of Alertmanager – users with two-factor email authentication may receive additional instructions for enabling notifications.

If everything worked well – congratulations, you have set up service alerting! We intentionally haven't covered many of Alertmanager's features – it includes many configurable settings that are outside the scope of this chapter. If you are interested in learning more about it, check the official documentation at `https://prometheus.io/docs`.

Now, let's proceed to the next section, where we will provide some best practices for setting up service alerting that should help you increase your service reliability.

Alerting best practices

The knowledge you will gain by reading this section should be useful for establishing the new alerting process for your services. It will also help you improve existing alerts if you are working with some established alerting processes.

Among the most valuable best practices, I would highlight the following ones:

- **Keep your alerts immediately actionable**: Alerting is a powerful technique for ensuring any issues or incidents get acknowledged and addressed. However, you should not overuse it for the types of issues that do not require immediate attention. Some types of alerts, such as alerts indicating high saturation, are not necessarily actionable. For example, a sudden increase in CPU load may not indicate any immediately actionable issue, unless it remains high for some prolonged period (for example, the CPU load not going below 85% for more than 10 minutes), and may just be a transient symptom of high service usage. When creating alerts, think about whether an engineer needs to perform any manual action as a result of a notification, and reduce any possible noise as much as possible (for example, specify how long the metric should be breaching the threshold before an alert gets fired by providing the `for` value in the rule configuration).

- **Include the runbook references**: For each alert, ensure you have a runbook in place that provides clear instructions to the on-call engineers receiving it. Having an accurate and up-to-date runbook for each alert helps reduce the incident mitigation time and share relevant knowledge among all engineers.

- **Ensure the alerting configuration is reviewed periodically**: The best solution for ensuring the alerting configuration is accurate is to make it easy to access and review. One of the easiest solutions is to make the alerting configuration a part of your code base so that all alert configurations are easily reviewable. Perform periodic checks of your alerts to ensure all important scenarios are covered, as well as to ensure no alerts are outdated.

This list contains just a handful of best practices to improve your service alerting. If you are interested in the topic, I strongly suggest that you read the relevant chapters of the *Google SRE* book, including the *Monitoring Distributed Systems* chapter from `https://sre.google/sre-book/monitoring-distributed-systems/`.

This summarizes a brief overview of service alerting. Now, let's summarize this chapter.

Summary

In this chapter, we covered one of the most important aspects of service reliability work – alerting. You learned how to set up the service metric collection using the Prometheus tool and the `tally` library, set up service alerts using the Alertmanager tool, and connect all these components to create an end-to-end service alerting pipeline.

The material in this chapter summarizes our learning from the reliability and service telemetry topics from *Chapter 10* and *Chapter 11*. By collecting the telemetry data and establishing the notification mechanisms using the alerting tools, we can quickly detect various service issues and get notified each time we need to mitigate them.

In the next chapter, we will continue covering some advanced aspects of Go development, including system profiling and dashboarding.

Further reading

To learn more about the topics that were covered in this chapter, take a look at the following resources:

- *Practical Alerting from Time-Series Data*: `https://sre.google/sre-book/practical-alerting/`

- *Monitoring Distributed Systems*: `https://sre.google/sre-book/monitoring-distributed-systems/`

- Prometheus documentation: `https://prometheus.io/docs/introduction/overview/`

- *Eliminating Toil*: `https://sre.google/workbook/eliminating-toil/`

13

Advanced Topics

If you are reading this chapter – congratulations, you have reached the very final part of this book! We have discussed many topics related to microservice development, but some remain that are important to cover. The topics in this chapter span many areas, from observability and debugging to service ownership and security. You may find these topics useful at various points in time: some of them will be helpful once you have working services serving production traffic, while others will be useful while your services are still in active development.

In this chapter, we will cover the following topics:

- Profiling Go services

- Creating microservice dashboards

- Frameworks

- Storing microservice ownership data

- Securing microservice communication with JWT

Let's proceed to the first section of this chapter, which covers service profiling.

Technical requirements

To complete this chapter, you will need Go 1.11+ or above. Additionally, you will need the following tools:

- **Graphviz**: `https://graphviz.org`

- **Docker**: `https://www.docker.com`

You can find the code examples for this chapter on GitHub: `https://github.com/PacktPublishing/microservices-with-go/tree/main/Chapter13`.

Profiling Go services

In this section, we are going to review a technique called **profiling**, which involves collecting real-time performance data of a running process, such as a Go service. Profiling is a powerful technique that can help you analyze various types of service performance data:

- **CPU usage**: Which operations used the most CPU power and what was the distribution of CPU usage among them?

- **Heap allocation**: Which operations used heap (dynamic memory allocated in Go applications) and what amount of memory was used?

- **Call graph**: In which order were service functions executed?

Profiling may help you in different situations:

- **Identifying CPU-intensive logic**: At some point, you may notice that your service is consuming most of your CPU power. To understand this problem, you can collect the CPU profile – a graph showing the CPU usage of various service components, such as individual functions. Components that consume too much CPU power may indicate various issues, such as inefficient implementations or code bugs.

- **Capturing the service memory footprint**: Similar to high CPU consumption, your service may be using too much memory (for example, to allocate too much data to the heap), resulting in occasional service crashes due to out-of-memory panics. Performing memory profiling may help you analyze the memory usage of various parts of your service and find components that have unexpectedly high memory usage.

Let's illustrate how to profile Go services using the pprof tool, which is a part of the Go SDK. To visualize the results of the tool, you will need to install the Graphviz library: `https://graphviz.org/`.

We will use the metadata service that we implemented in *Chapter 2* as an example. Open the `metadata/cmd/main.go` file and add the `flag` package to the `imports` block. Then, add the following code to the beginning of the main function, immediately after the logger initialization:

```
simulateCPULoad := flag.Bool("simulatecpuload",
    false,"simulate CPU load for profiling")
    flag.Parse()
    if *simulateCPULoad {
        go heavyOperation()
    }


go func() {
    if err := http.ListenAndServe("localhost:6060", nil);
```

```
        err != nil {
            logger.Fatal("Failed to start profiler handler",
                zap.Error(err))
        }
    }()
```

In the code we just added, we introduced an additional flag called `simulatecpuload` that will let us simulate a CPU-intensive operation for our profiling. We also started an HTTP handler that we will use to access the profiler data from the command line.

Now, let's add another function to the same file that will run a continuous loop and execute some CPU-intensive operations. We will generate random 1,024-byte arrays and calculate their md5 hashes (you can read about the md5 operation in the comments of its Go package at `https://pkg.go.dev/crypto/md5`). Our selection of such logic is fully arbitrary: we could easily choose any other operation that would consume some visible part of the CPU load.

Add the following code to the `main.go` file that we just updated:

```
func heavyOperation() {
    for {
        token := make([]byte, 1024)
        rand.Read(token)
        md5.New().Write(token)
    }
}
```

Now, we are ready to test our profiling logic. Run the service with the `--simulatecpuload` argument:

```
go run *.go --simulatecpuload
```

Now, execute the following command:

```
go tool pprof http://localhost:6060/debug/pprof/
profile?seconds=5
```

The command should take 5 seconds to complete. If it executes successfully, the pprof tool will be running, as shown here:

```
Type: cpu
Time: Sep 13, 2022 at 5:37pm (+05)
Duration: 5.14s, Total samples = 4.42s (85.92%)
Entering interactive mode (type "help" for commands,
    "o" for options)
(pprof)
```

Type web in the command prompt of the tool and press *Enter*. If everything worked well, you will be redirected to a browser window containing a CPU profile graph:

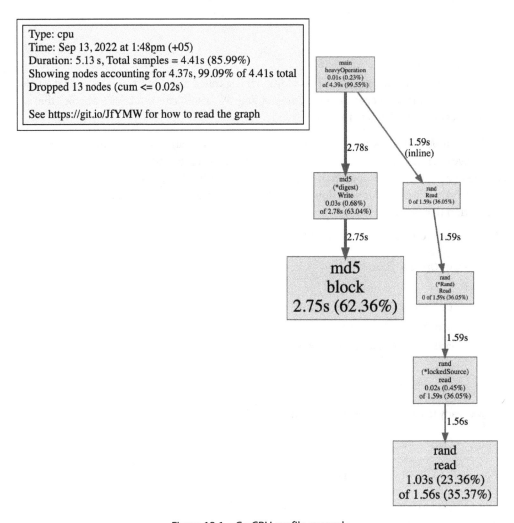

Figure 13.1 – Go CPU profile example

Let's walk through the data from the graph to understand how to interpret it. Each node on the graph includes the following data:

- Package name

- Function name

- Elapsed time and the total time of the execution

For example, the `heavyOperation` function took just 0.01 seconds, but all the operations that were executed in it (including all function calls inside it) took 4.39 seconds, taking most of the elapsed time.

If you walk through the graph, you will see the distribution of the elapsed time by sub-operations. In our case, `heavyOperation` executed two functions that got recorded by the CPU profiler: `md5.Write` and `rand.Read`. The `md5.Write` function took 2.78 seconds in total, while `rand.Read` took 1.59 seconds of the execution time. Level by level, you can analyze the calls and find the CPU-intensive functions.

When working with the CPU profiler data, notice the functions that take the most processing time. Such functions are illustrated as larger rectangles to help you find them. If you notice that some functions have unexpectedly high processing time, spend some time analyzing their code to see whether there is any opportunity to optimize them.

Now, let's illustrate another example of profiler data. This time, we will be capturing a **heap profile** – a profile showing dynamic memory allocation by a Go process. Run the following command:

```
go tool pprof http://localhost:6060/debug/pprof/heap
```

Similar to the previous example, successfully executing this command should run the pprof tool, where we can execute a web command. The result will contain the following graph:

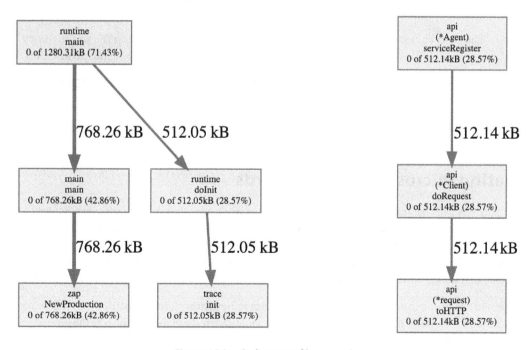

Figure 13.2 – Go heap profile example

This diagram is similar to the CPU profile. The last line inside each node shows the ratio between the memory used by the function and the total heap memory allocated by the process.

In our example, three high-level operations are consuming the heap memory:

- **api.serviceRegister**: A function that registers a service via the Consul API

- **zap.NewProduction**: Logger initialization via the `zap` library

- **trace.init**: Initializes the tracing logic

1. Looking at the heap profiler data, it's easy to find functions allocating an unexpectedly high amount of heap memory. Similar to CPU profiler graphs, heap profilers display the functions that have the highest heap allocation as larger rectangles, making it easier to visualize the most memory-consuming functions.

2. I suggest that you practice with the pprof tool and try the other operations it provides. Being able to profile Go applications is a highly valuable skill in production debugging that should help you optimize your services and solve different performance-related issues. The following are some other useful tips for profiling Go services:

   - You can profile Go tests without adding any extra logic to your code. Running the `go test` command with the `-cpuprofile` and `-memprofile` flags will capture the CPU and memory profiles of your logic, respectively.

   - The `top` command of the pprof tool is a convenient way of showing the top memory consumers. There is also the `top10` command, which shows the top 10 memory consumers.

   - Using the `roroutine` mode of the pprof tool, you can get a profile of all used goroutines, as well as their stack traces.

Now that we have covered the basics of Go profiling, let's move on to the next topic of this chapter: service dashboarding.

Creating microservice dashboards

In the previous two chapters, we reviewed various ways of working with service metrics. In *Chapter 11*, we demonstrated how to collect the service metrics, while in *Chapter 12*, we showed you how to aggregate and query them using the Prometheus tool. In this section, we will describe one more way of accessing the metrics data that can help you explore your metrics and plot them as charts. The technique that we will cover is called **dashboarding** and is useful for visualizing various service metrics.

Let's provide an example of a dashboard – a set of charts representing different metrics. The following figure shows the dashboard of a Go service containing some system-level metrics, such as the goroutine count, the number of Go threads, and allocated memory size:

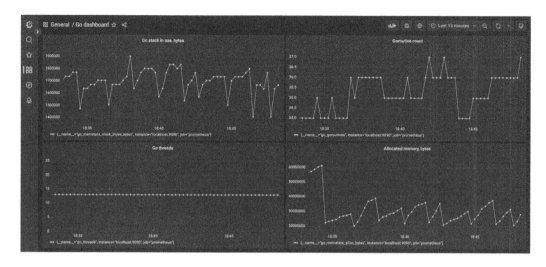

Figure 13.3 – Go process dashboard example from the Grafana tool

Dashboards help visualize various types of data, such as time series datasets, allowing us to analyze service performance. The following are some other use cases for using dashboards:

- **Debugging**: Being able to visualize various service performance metrics helps us identify service issues and notice any anomalies in system activity

- **Data correlation**: Having a side-by-side representation of multiple service performance charts helps us find related events, such as an increase in server errors or a sudden drop in available memory

It's a great practice to have a dashboard for each of your services, as well as some dashboards that span all services, to get some high-level system performance data, such as the number of active service instances, network throughput, and much more.

Let's demonstrate how to set up an example dashboard for the Prometheus data that we collected in *Chapter 12*. For this, we will use the open source tool called Grafana, which has built-in support for various types of time series data and provides a convenient user interface for setting up different dashboards. Follow these instructions to set up a Grafana dashboard:

1. Execute the following command to run the Grafana Docker image:

    ```
    docker run -d -p 3000:3000 grafana/grafana-oss
    ```

    This command should fetch and run the open source version of Grafana (Grafana also comes in an enterprise version, which we won't cover in this chapter) and expose port 3000 so that we can access it via HTTP.

> **Note**
>
> Similar to Prometheus, Grafana is also written in Go and is another example of a popular open source Go project widely used across the software development industry.

2. Once you've run the preceding command, open `http://localhost:3000` in your browser. This will lead you to the Grafana login page. By default, the Docker-based version of Grafana includes a user with `admin` as both its username and password, so you can use these credentials to log in.

3. From the side menu, select **Configuration**:

Figure 13.4 – Grafana data source configuration menu

4. On the **Configuration** page, click on the **Data sources** menu item, then click **Add data source** and choose **Prometheus** from the list of available data sources. Doing so will open a new page that displays Prometheus settings. In the **HTTP** section, set **URL** to `http://host.docker.internal:9090`, as shown in the following screenshot:

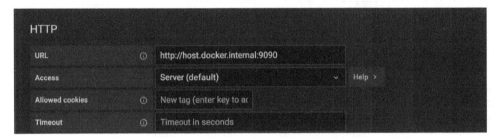

Figure 13.5 – Grafana configuration for a Prometheus data source

5. Now, you can click the **Save and test** button at the bottom of the page, which should let you know whether the operation was successful. If you did everything well, Grafana should be ready to display your metrics from Prometheus.

6. From the side menu, click on **New dashboard**:

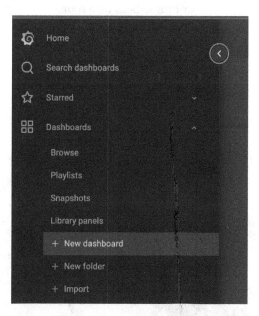

Figure 13.6 – Grafana's New dashboard menu item for dashboard creation

7. This should open an empty dashboard page.

8. Click on the **Add a new panel** button on this dashboard page; you will be redirected to the panel creation page.

A panel is a core element of a Grafana dashboard and its purpose is to visualize the provided dataset. To illustrate how to use it, let's select our Prometheus data source and some of the metrics that it already has. On the panel view, choose **Prometheus** as the data source and, in the **Metric** field, find the `process_open_fds` element and select it. Now, click on the **Run queries** button; you should see the following view:

Figure 13.7 – Grafana panel view

We just configured the dashboard panel to display the `process_open_fds` time series stored in Prometheus. Each data point on the chart shows the value of the time series at a different time, displayed below the chart. On the right-hand panel, you can set the panel title to **Open fd count**. Now, save the dashboard by clicking the **Apply** button provided in the top menu. You will be redirected to the dashboard page.

In the top menu, you will find the **Add panel** button, which you can use to add a new panel to our dashboard. If you follow the same steps that we did for the previous panel and choose the `go_gc_duration_seconds` metric, you will add a new panel to the dashboard that will visualize the `go_gc_duration_seconds` time series from Prometheus.

The resulting dashboard should look like this:

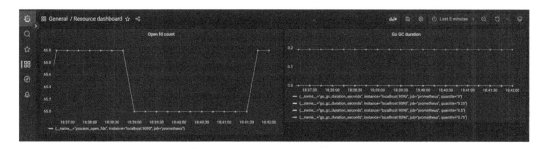

Figure 13.8 – Example Grafana dashboard

We just created an example dashboard that has two panels that display some existing Prometheus metrics. You can use the same approach to create any dashboards for your services, as well as high-level dashboards showing the system-global metrics, such as the total number of API requests, network throughput, or the total number of all service instances.

Let's provide some examples of metrics that can be useful for setting up a dashboard for an individual service. This includes *The Four Golden Signals*, which we mentioned in *Chapter 12*:

- **Client error rate**: The ratio between client errors (such as invalid or unauthenticated requests) and all requests to the service

- **Server error rate**: The ratio between server errors (such as database write errors) and all requests to the service

- **API throughput**: Number of API requests per second/minute

- **API latency**: API request processing latency, usually measured in percentiles, such as p90/p95/ p99 (you can learn about percentiles by reading this blog post: `https://www.elastic. co/blog/averages-can-dangerous-use-percentile`)

- **CPU utilization**: Current usage of CPUs (100% means all CPUs are fully loaded)

- **Memory utilization**: Ratio between used and total memory across all service instances

- **Network throughput**: Total amount of network write/read traffic per second/minute

Depending on the operations performed by your service (for example, database writes or reads, cache usage, Kafka consumption, or production), you may wish to include additional panels that will help you visualize your service performance. Make sure that you cover all the high-level functionality of the service so that you can visually notice any service malfunctions on your dashboards.

The Grafana tool, which we used in our example, also supports lots of different visualization options, such as displaying tables, heatmaps, numerical values, and much more. We will not cover these features in this chapter, but you can get familiar with them by reading the official documentation: `https://grafana.com/docs/`. Using the full power of Grafana will help you set up excellent dashboards for your services, simplifying your debugging and performance analysis.

Now, let's move on to the next section, where we will describe Go frameworks.

Frameworks

In *Chapter 2*, we covered the topic of the Go project structure, as well as some common patterns of organizing your Go code. The code organization principles that we described are generally based on conventions – written agreements or statements that define specific rules for naming and placing Go files. Some of the conventions that we followed were proposed by the authors of the Go language, while others are commonly used and proposed by authors of various Go libraries.

While conventions play an important role in establishing the common principles of organizing Go code, there are other ways of enforcing specific code structures. One such way is using frameworks, which we are going to cover in this section.

Generally speaking, **frameworks** are tools that establish a structure for various components of your code. Let's take the following code snippet as an example:

```go
package main

import (
    "fmt"
    "net/http"
)

func main() {
    http.HandleFunc("/echo",
        func(w http.ResponseWriter, _ *http.Request) {
            fmt.Fprintf(w, "Hi!")
        })
    if err := http.ListenAndServe(":8080", nil);
    err != nil {
        panic(err)
    }
}
```

Here, we are registering an HTTP handler function and letting it handle HTTP requests on the `localhost:8080/echo` endpoint. The code for our example is extremely simple, yet it does a lot of background work (you can check the source of the `net/http` package to see how complex the internal part of the HTTP handling logic is) to start an HTTP server, accept all incoming requests, and respond to them by executing the function provided by us. Most importantly, our code allows us to add additional HTTP handlers by following the same format of calling the `http.HandleFunc` function and passing handler functions to it. The `net/http` library that we used in our example established a structure for handling HTTP calls to various endpoints, acting as a framework for our Go application.

The authors of the `net/http` package were able to add additional HTTP endpoint handlers (provided by the `http.HandleFunc` function) by following a pattern called **inversion of control** (**IoC**). IoC is a way of organizing code in which some component (in our case, the `net/http` package) takes control of the execution flow by calling the other components of it (in our case, the function provided as an argument to `http.HandleFunc`). In our example, the moment we call the `http.ListenAndServe` function, the `net/http` package takes control of executing the HTTP handler functions: each time the HTTP server receives an incoming request, our function is called automatically.

IaC is a primary mechanism of most frameworks that allows them to establish a foundation for various parts of application code. In general, most frameworks work by taking control of an application, or a part of it, and handling some routing operations, such as resource management (opening and closing incoming connections, writing and reading files, and so on), serialization and deserialization, and many more.

What are the primary use cases for using Go frameworks? We can list some of the most common ones:

- **Writing web servers**: Similar to our example of an HTTP server, there can be other types of web servers handling requests to different endpoints using different protocols, such as Apache Thrift or gRPC.

- **Async event processing**: There are libraries for various asynchronous communication tools, such as Apache Kafka, that help organize code in an IoC way by passing handler functions for various types of events (such as Kafka messages belonging to different topics), which get called automatically each time there is a new unprocessed message.

It is important to note that frameworks have some significant downsides:

- **Harder to debug and understand the execution flow**: In addition to taking control of the execution flow, frameworks also perform lots of background work that is hidden from developers. Because of this, it is usually much harder to understand how your code is being executed, as well as to debug various issues, such as initialization errors (you may find more information on this in the following article: `https://www.baeldung.com/cs/framework-vs-library`).

- **Steeper learning curve**: Frameworks generally require a good understanding of the logic and abstractions they provide. This requires developers to spend more time reading the related documentation or learning some key lessons in practice.

- **Harder to catch some trivial bugs via static checks**: Frameworks often use dynamic code invocation libraries, such as `reflect` (`https://pkg.go.dev/reflect`). Such operations are performed when executing a program, making it hard to catch various types of issues, such as the incorrect implementation of interfaces or invalid naming.

When deciding on using a specific framework, you should do some analysis and compare the advantages it provides to you with the downsides it brings, especially in the long term. Many developers underestimate the complexity that frameworks bring to them or the other developers in their organizations: most frameworks perform a fair amount of *magic* to provide a convenient code structure to application developers. In general, you should always start from a simpler option (in our case, not using a particular framework) and only decide to use a framework if its benefits outweigh its downsides.

Now that we have discussed the topic of frameworks, let's move on to the next section, where we will describe the different aspects of microservice ownership.

Storing microservice ownership data

One of the key benefits of using microservice architectures is the ability to distribute their development: each service can be developed and maintained by a separate team, and teams can be distributed across the globe. While the distributed development model helps different teams build various parts of their systems independently, it brings some new challenges, such as service ownership.

To illustrate the problem of service ownership, imagine that you are working in a company with thousands of microservices. One day, the security engineers of your company find out that there is a critical security vulnerability in a popular Go library that is used in most of the company's services. How can you communicate with the right teams and find out who would be responsible for making the changes in each service?

There are numerous companies with thousands of microservices. In such companies, it becomes impossible to remember which team and which developers are responsible for each of them. In such companies, it becomes crucial to find a solution to the service ownership problem.

> **Note**
> While we are discussing the ownership problem for microservices, the same principles apply to many other types of technological assets, such as Kafka topics and database tables.

How can we define service ownership?

There are many different ways of doing this, each of which is equally important for some specific use cases:

- **Accountability**: Which person/entity is accountable for the service and who can act as the primary point of contact or the main authority for it?

- **Support**: Who is going to provide support for the service, such as a service bug, feature request, or user question?

- **On-call**: Who is currently on-call for the service? Who can we contact in case of an emergency issue?

As you can see, there are many ways of interpreting the word ownership, depending on the use case. Let's look at some ways to define each role, starting with accountability: who should be accountable, or liable, for a service?

In most organizations, liability is attributed to engineering managers: every engineering manager acts as an accountable individual for some unique domain. If you define a mapping between your services and the engineering managers that are responsible for them, you can solve the service accountability problem by allowing them to easily find the relevant point of contact, such as an engineering manager that is liable for it.

An alternative way of defining service accountability is to associate services with teams. However, there can be multiple issues with this:

- **Shared accountability does not always work**: If you have multiple people that are responsible for a service, it becomes unclear who the final authority among them is.

- **A team is a loosely defined concept in many organizations**: Unless you have a single, well-defined registry of teams in your company, it's better to avoid referencing team names in your systems.

Now, let's discuss the support aspect of ownership. Ideally, each service should have a mechanism for reporting any issues or bugs. Such a mechanism can take one of the following forms:

- **Support channel**: The identifier or URL of a messaging channel for leaving support requests, such as a link to the relevant Google group, Slack channel, or any other similar tool.

- **Ticketing system URL**: The URL to a system/page that allows you to create a support request ticket. Developers often use Atlassian Jira for this purpose.

If you provide such metadata for all your services, you will significantly simplify user support: all service users, such as other developers, will always know how to request support for them or report any bugs or other issues.

Let's move on to the on-call ownership metadata. An easy solution to this is to link each service to its on-call rotation. If you use PagerDuty, you can store the relationships between service names and their corresponding PagerDuty rotation identifiers.

An example of the ownership metadata that we just described is as follows:

```
ownership:
    rating-service:
        accountable: example@somecompany.com
            support:
                slack: rating-service-support-group
                    oncall:
                        pagerduty_rotation:SOME_ROTATION_ID
```

Our example is defined in YAML format, though it may be preferable to store this data in some system that would allow us to query or modify it via an API. This way, you can automatically submit new ownership changes (for example, when people leave the company and you want to reassign the ownership automatically). I would also suggest making the ownership data mandatory for all services. To enforce this, you can establish a service creation process that will request the ownership data before developers provision new services.

Now that we've discussed service ownership, let's move on to the next section, where we will describe the basics of Go microservice security.

Securing microservice communication with JWT

In this section, we are going to review some basic concepts of microservice security, such as authentication and authorization. You will learn how to implement such logic in Go using a popular **JSON Web Token (JWT)** protocol.

Let's start with one of the primary aspects of security: **authentication**. Authentication is the process of verifying someone's identity, such as via user credentials. When you log into some system, such as Gmail, you generally go through the authentication process by providing your login details (username and password). The system that performs authentication performs verification by comparing the provided data with the existing records it stores. Verification can take one or multiple steps: some types of authentication, such as two-factor authentication, require some additional actions, such as verifying access to a phone number via SMS.

A successful authentication often results in granting the caller access to some resources, such as user data (for example, user emails in Gmail). Additionally, the server performing this authentication may provide a security token to the caller that can be used on subsequent calls to skip the verification process.

Another form of access control, known as **authorization**, involves specifying access rights to various resources. Authorization is often performed to check whether a user has permission to perform a certain action, such as viewing a specific admin page. Authorization is often performed by using a security token that was obtained during authentication, as illustrated in the following diagram:

Figure 13.9 – Authorization request providing a token

There are many different ways to implement authentication and authorization in microservices. Among the most popular protocols is JWT, a proposed internet standard for creating security tokens that can contain any number of facts about the caller's identity, such as them being an administrator. Let's review the basics of the protocol to help you understand how to use it in your services.

JWT basics

JWTs are generated by components that perform authentication or authorization. Each token consists of three parts: a header, a payload, and a signature. The payload is the main part of the token and it contains a set of **claims** – statements about the caller's identity, such as a user identifier or a role in the system. The following code shows an example of a token payload:

```
{
    "name": "Alexander",
    "role": "admin",
    "iat": 1663880774
}
```

Our example payload contains three claims: the user's name, role (admin in our example), and token issuance time (iat is a standard field name that is a part of the JWT protocol). Such claims could be used in various flows – for example, when checking whether a user has the admin role to access a system dashboard.

As a protection mechanism against modifications, each token contains a **signature** – a cryptographic function of its payload - a header, and a special value, called a **secret**, that is known only to the authentication server. The following pseudocode provides an example of token signature calculation:

```
HMACSHA256(
    base64UrlEncode(header) + "." +
    base64UrlEncode(payload),
```

```
        secret,
    )
```

The algorithm that is used for creating a token signature is defined in a **token header**. The following JSON record provides an example of a header:

```
    {
        "alg": "HS256",
        "typ": "JWT"
    }
```

In our example, the token is using HMAC-SHA256, a cryptographic algorithm that is commonly used for signing JWTs. Our selection of HMAC-SHA256 is primarily due to its popularity; if you wish to learn about other signing algorithms, you can find a link to an overview of them in the *Further reading* section of the chapter.

The resulting JWT is a concatenation of the token's header, payload, and signature, encoded with the *Base64uri* protocol. For example, the following value is a JWT that's been created by combining the header and the payload from our code snippets, signed with a secret string called our-secret:

```
    eyJhbGciOiJIUzI1NiIsInR5cCI6IkpXVCJ9.eyJuYW1lIjoiQWxleGFuZGVy-
    Iiwicm9sZSI6ImFkbWluIiwiaWF0IjoxNjYzODgwNzc0fQ.FqogLyrV28wR-
    5po6SMouJ7qs2Y3m6gmpaPg6MUthWpQ
```

To practice JWT creation, I suggest using the JWT tool available at https://jwt.io to try encoding arbitrary JWTs and see the resulting token values.

Now that we have discussed the high-level details of JWT, let's move on to the practical part of this section – implementing basic authentication and authorization in Go microservices using JWTs.

Implementing authentication and authorization with JWTs

In this section, we will provide some examples of implementing basic access control via authentication and authorization using Go.

Let's start with the authentication process. A simple credential-based authentication flow can be summarized as follows:

- The client initiating authentication would call a specified endpoint (for example, HTTPS POST /auth) while providing the user credentials, such as username and password.
- The server handling authentication would verify the credentials and perform one of two actions:

- Return an error if the credentials are invalid (for example, an HTTP error with a 401 code).

- Return a successful response with a 200 code, containing a JWT, that is signed with the server's secret.

- If authentication is successful, the client can store the received token so that it can be used in the following requests.

Let's illustrate how to implement the server logic for the authentication flow that we just described. To generate JWTs in our Go code, we will use the https://github.com/golang-jwt/jwt library.

The following code provides an example of handling an HTTP authentication request. It performs credential validation and returns a successful response with a signed JWT if the validation passes:

```go
const secret = "our-secret"


func Authenticate(w http.ResponseWriter, req *http.Request) {
    username := req.FormValue("username")
    password := req.FormValue("password")
    if !validCredentials(username, password) {
        http.Error(w, "invalid credentials", http.
StatusUnauthorized)
        return
    }
    token := jwt.NewWithClaims(jwt.SigningMethodHS256, jwt.
MapClaims{
        "username": username,
        "iat": time.Now().Unix(),
    })
    tokenString, err := token.SignedString(secret)
    if err != nil {
        http.Error(w, "failed to create a token", http.
StatusInternalServerError)
        return
    }
    fmt.Fprintf(w, tokenString)
}


func validCredentials(username, password string) bool {
    // Implement your credential verification here.
```

```
        return false
}
```

In the preceding code, we created a token using the `jwt.NewWithClaims` function. The token includes two fields:

- `username`: Name of the authenticated user

- `iat`: Time of token creation

The server code that we just created is using the secret value to sign the token. Any attempts to modify the token would be impossible without knowing the secret: the token signature allows us to check whether the token is correct.

Now, let's illustrate how the client can perform requests using the token that it receives after successfully authenticating:

```
func authorizationExample(
    token string, operationURL string) error {
        req, err := http.NewRequest(
            http.MethodPost, operationURL, nil)
            if err != nil {
                return err
            }
        req.Header.Set("Authorization", "Bearer "+token)
        resp, err := http.DefaultClient.Do(req)
        // Handle response.
}
```

In our example of an authorized operation, we added an `Authorization` header to the request while using the token value with the `Bearer` prefix. The `Bearer` prefix defines a **bearer token** – a token that intends to give access to its bearer.

Let's also provide the logic of a server handler that would handle such an authorized request and verify whether the provided token is correct:

```
func AuthorizedOperationExample(w http.ResponseWriter,
    req *http.Request) {
        authHeaderValue := req.Header.Get("Authorization")
        const bearerPrefix = "Bearer "
        if !strings.HasPrefix(authHeaderValue,
            bearerPrefix) {
```

```go
        http.Error(w,
            "request does not contain an Authorization
Bearer token", http.StatusUnauthorized)
            return
        }
    tokenString := strings.TrimPrefix(authHeaderValue,
        bearerPrefix)
    // Validate token.
    token, err := jwt.Parse(tokenString,
        func(token *jwt.Token) (interface{}, error) {
            if _, ok := token.Method.(
                *jwt.SigningMethodHMAC); !ok {
                    return nil,
                    fmt.Errorf(
                        "unexpected signing method:
                            %v", token.Header["alg"])
            }
            return secret, nil
    })
    if err != nil {
        http.Error(w, "invalid token",
            http.StatusUnauthorized)
    }

    claims, ok := token.Claims.(jwt.MapClaims)
    if !ok || !token.Valid {
        http.Error(w, "invalid token",
            http.StatusUnauthorized)
        return
    }
    username := claims["username"]
    fmt.Fprintf(w, "Hello, "+username.(string))
}
```

Let's describe some highlights of the provided example:

- We use the `jwt.Parse` function to parse the token and validate it. We return an error if the signature algorithm does not match `HMAC-SHA256`, which we used previously.

- The parsed token contains the `Claims` field, which contains the claims from the token payload

- We use the `username` claim from the token payload in our function. Once we successfully parse the token and verify that it is valid, we can assume that the information in its payload has been securely passed to us and can trust it.

Now that we have provided examples of Go authentication and authorization using JWTs, let's list some best practices for using JWTs to secure microservice communication:

- **Set a token expiration time**: When issuing JWTs, it is useful to set the token expiration time (the `exp` JWT claim field) to avoid situations where users use old authorization records. By having an expiration time set in each token payload, you can verify it against authorization requests. For example, when a user authenticates as a system administrator, you can set a short token expiration time (for example, a few hours) to avoid situations where a former administrator can still perform critical actions in the system.

- **Include the token issuance time**: Additional metadata, such as the token issuance time (the `ist` JWT claim field), can be useful in many practical situations. For example, if you identify a security breach that happened at a certain point in time, you can invalidate all access tokens that were issued before that moment by using the token issuance time metadata.

- **Use JWTs with HTTPS instead of HTTP**: The HTTPS protocol encrypts request metadata, such as authorization request headers, preventing various types of security attacks. An example of such a security attack is a *man-in-the-middle* attack, which is when some third party (such as a hacker trying to obtain a user's access token) captures network traffic to extract JWTs from request headers.

- **Prefer standard JWT claim fields to custom ones**: When including metadata in the JWT payload, make sure that there is no standard field for the same purpose. You can find a list of standard JWT claim fields at `https://en.wikipedia.org/wiki/JSON_Web_Token#Standard_fields`.

The `https://jwt.io/` website contains some additional tips on using JWTs, as well as an online tool for encoding and decoding JWTs, that you can use to debug your service communication.

Summary

With that, we have finished the last chapter of this book by reviewing lots of microservice development topics that were not included in the previous chapters. You learned how to profile Go services, create microservice dashboards so that you can monitor their performance, define and store microservice ownership data, and secure microservice communication with JWTs. I hope that you have found lots of interesting tips in this chapter that will help you build scalable, highly performant, and secure microservices.

The Go language keeps evolving, as well as the tooling for it. Each day, developers release new libraries and tools for it that can solve various microservice development problems that we described in this book. While this book provided you with lots of tips on Go microservice development, you should keep improving your skills and make your services simpler and easier to maintain.

I also want to thank you for reading this book. I hope you enjoyed reading it and gained lots of useful experience that will help you in mastering the art of Go microservice development. Let your Go microservices be highly performant, secure, and easy to maintain!

Further reading

To learn more about the topics that were covered in this chapter, take a look at the following resources:

- *Profiling Go Programs*: `https://go.dev/blog/pprof`
- Grafana documentation: `https://grafana.com/docs/`
- *Grafana support for Prometheus*: `https://prometheus.io/docs/visualization/grafana/`
- Top Go frameworks: `https://github.com/mingrammer/go-web-framework-stars`
- JSON web token: `https://jwt.io/`
- *JSON Web Token (JWT) Signing Algorithms Overview*: `https://auth0.com/blog/json-web-token-signing-algorithms-overview/`

We suggest that you use these resources to stay up to date with the latest news related to Go microservice development:

- **The Go blog**: `https://go.dev/blog/`
- **Microservice architecture**: `https://microservices.io`
- **A curated list of awesome Go software**: `https://github.com/avelino/awesome-go`

Index

H

HashiCorp Consul 60
　URL 60
high-cardinality data 236
histograms 230
HTTP protocol
　client error 94
　headers 94
　request body 94
　response body 94
　server error 94
　URL parameters 94

I

idiomatic Go code
　comments 16
　errors 16
　naming 15
　writing 15
incident management 210
incident postmortems 211
in-memory service discovery
　　implementation 64-67
integration test 175, 176
　challenges, with using existing
　　persistent databases 186
　implementing 177-186
　plan, writing for 177
　structure 176
　writing suggestions 187
interceptors 203
internal model 100

J

Jaeger 239
Jenkins
　URL 187
jittering 199
JSON 80
　benefits 80
　limitations 80
JSON Web Token (JWT)
　authentication, implementing with 284-286
　authorization, implementing with 286-288
　basics 283, 284
　best practices, for securing microservice
　　communication 288

K

key-value databases 136
Kibana 224
Kubernetes 60
　benefits 153
Kubernetes deployments
　microservices, setting up for 153-158

L

labels 232
latency tracking 231
log 217
logging 217
logging library
　features 220
log levels
　debug 219
　error 219
　fatal 219

`Packt.com`

Subscribe to our online digital library for full access to over 7,000 books and videos, as well as industry leading tools to help you plan your personal development and advance your career. For more information, please visit our website.

Why subscribe?

- Spend less time learning and more time coding with practical eBooks and Videos from over 4,000 industry professionals

- Improve your learning with Skill Plans built especially for you

- Get a free eBook or video every month

- Fully searchable for easy access to vital information

- Copy and paste, print, and bookmark content

Did you know that Packt offers eBook versions of every book published, with PDF and ePub files available? You can upgrade to the eBook version at `packt.com` and as a print book customer, you are entitled to a discount on the eBook copy. Get in touch with us at `customercare@packtpub.com` for more details.

At `www.packt.com`, you can also read a collection of free technical articles, sign up for a range of free newsletters, and receive exclusive discounts and offers on Packt books and eBooks.

Other Books You May Enjoy

If you enjoyed this book, you may be interested in these other books by Packt:

Go for DevOps

John Doak, David Justice

ISBN: 978-1-801-81889-6

- Understand the basic structure of the Go language to begin your DevOps journey
- Interact with filesystems to read or stream data
- Communicate with remote services via REST and gRPC
- Explore writing tools that can be used in the DevOps environment
- Develop command-line operational software in Go
- Work with popular frameworks to deploy production software
- Create GitHub actions that streamline your CI/CD process
- Write a ChatOps application with Slack to simplify production visibility

Event-Driven Architecture in Golang

Michael Stack

ISBN: 978-1-803-23801-2

- Learn different event-driven patterns and best practices.
- Plan and design your software architecture with ease.
- Effectively track changes and updates using event sourcing.
- Test and deploy your sample software application with ease.
- Monitor and improve the performance of your software architecture.

Packt is searching for authors like you

If you're interested in becoming an author for Packt, please visit `authors.packtpub.com` and apply today. We have worked with thousands of developers and tech professionals, just like you, to help them share their insight with the global tech community. You can make a general application, apply for a specific hot topic that we are recruiting an author for, or submit your own idea.

Share Your Thoughts

Now you've finished *Microservices with Go*, we'd love to hear your thoughts! Scan the QR code below to go straight to the Amazon review page for this book and share your feedback or leave a review on the site that you purchased it from.

https://packt.link/r/1804617008

Your review is important to us and the tech community and will help us make sure we're delivering excellent quality content.

Download a free PDF copy of this book

Thanks for purchasing this book!

Do you like to read on the go but are unable to carry your print books everywhere?

Is your eBook purchase not compatible with the device of your choice?

Don't worry, now with every Packt book you get a DRM-free PDF version of that book at no cost.

Read anywhere, any place, on any device. Search, copy, and paste code from your favorite technical books directly into your application.

The perks don't stop there, you can get exclusive access to discounts, newsletters, and great free content in your inbox daily

Follow these simple steps to get the benefits:

1. Scan the QR code or visit the link below

https://packt.link/free-ebook/978-1-80461-700-7

2. Submit your proof of purchase
3. That's it! We'll send your free PDF and other benefits to your email directly

www.ingramcontent.com/pod-product-compliance
Lightning Source LLC
Chambersburg PA
CBHW062103050326
40690CB00016B/3190